INTRODUCTION TO APPLIED MATHEMATICS FOR ENVIRONMENTAL SCIENCE

INTRODUCTION TO APPLIED MATHEMATICS FOR ENVIRONMENTAL SCIENCE

by

David F. Parkhurst
Indiana University
Bloomington, IN

 Springer

ISBN-13: 978-1-4419-4169-5   e-ISBN-10: 0-387-34228-1
                                    e-ISBN-13: 978-0-387-34228-3

Printed on acid-free paper.

MATLAB® is a registered trademark of The Math Works, Inc.

Printed in the United States of America.

9 8 7 6 5 4 3 2 1

springer.com

Preface

For many years, first as a student and later as a teacher, I have observed graduate students in ecology and other environmental sciences who had been required as undergraduates to take calculus courses. Those courses have often emphasized how to prove theorems about the beautiful, logical structure of calculus, but have neglected applications. Most of the time, the students have come out of such courses with little or no appreciation of how to *apply* calculus in their own work. Based on these observations, I developed a course designed in part to re-teach calculus as an everyday tool in ecology and other environmental sciences. I emphasized derivations—working with story problems (sometimes quite complex ones)—in that course, and now in this book.

The present textbook has developed out of my notes for that course. Its basic purpose is to describe various types of mathematical structures and how they can be applied in environmental science. Thus, linear and non-linear algebraic equations, derivatives and integrals, and ordinary and partial differential equations are the basic kinds of structures, or types of mathematical models, discussed. For each, the discussion follows a pattern something like this:

1. An example of the type of structure, as applied to environmental science, is given.

2. Next, a description of the structure is presented.

3. Usually, this is followed by other examples of how the structure arises in environmental science.

4. The analytic methods of solving and learning from the structure are discussed.

5. Numerical methods for use when the going gets too rough analytically are described.

6. In most chapters, examples of using MATLAB® software to solve and explore the structures are also included. All these examples have been tested with Version 7, Releases R14 and R2006a.

This book is *not* an introduction to calculus—it assumes that its readers will already have been introduced to the basic ideas of differential and integral calculus. It does, however, include three early chapters and an appendix to review basic algebra, derivatives, and integrals.

So far as I know, the combination of materials provided in the book is unique, but I believe it forms the basis for a useful and interesting course. In general, none of the material goes beyond what might be taught in a junior-level math or engineering course, but because the book covers ground from several such courses, the present material is appropriately taught at the graduate level. Obviously then, parts of the material treated here could be selected for use in an undergraduate course—indeed, advanced undergraduates have often done well in my version of the course.

In addition to its use as a text for a course, the material here should provide an interesting source for environmental scientists and managers to review forgotten math, and to learn some that is new.

Environmental science is a broad area, and I have included examples, and over 150 exercises, drawn from a wide variety of its subfields. A list of applications is provided in Appendix C.

In my classes, I asked students to write out questions at the end of each period, and then answered those to the whole class by e-mail. Selections of those questions and answers are provided at the end of most chapters.

Readers wishing a review of basic math may find Appendix A helpful. Over the nearly 30 years I've taught the course that led to this book, I've discovered that many students have apparently not learned to study math and other quantitative subjects effectively. For that reason, I recommend having a look at the study suggestions provided at the beginning of Appendix B, on p. 292.

I thank the many students and colleagues who have helped me tune these notes over the years. Special thanks go to Deborah Robinson for many useful suggestions and careful proofreading of the entire text. As always, any remaining errors are my responsibility.

Contents

Chapter 1

Introduction

1.1 On Translating Ideas to Mathematics

In a sense, this book is about how to work environmental science "story problems." It is often useful to solve such problems symbolically first; i.e., in terms of letter variables (a, b, x, y, etc.), and to put in numerical values only near the end of each problem. Consider an example:

> Your laboratory keeps two stock solutions of ethanol, one
> with 90% and one with 40% of alcohol in water. How much
> of each of these two solutions must be mixed to produce
> 1 liter of a solution that is 2/3 alcohol?

You could solve this problem numerically for the particular case involved, but if other stocks, or other final alcohol concentrations, might be needed in the future, it would be useful to solve the general case in terms of symbols. To do this:

- First define what you know *in terms of variables*, stating units for each.

For example, let

f_1 = fraction of alcohol in Solution 1 = 0.9 L alcohol/L solution

f_2 = fraction of alcohol in Solution 2 = 0.4 L alcohol/L solution

f_3 = fraction of alcohol in final solution = 2/3 L alcohol/L solution

V_3 = liters of final solution = 1 liter.

- Next write descriptions of quantities that you *don't* know. Again use symbols and give units.

V_1 = liters of Solution 1 needed = unknown

V_2 = liters of Solution 2 needed = unknown

- Many problems in environmental science involve mass balances or energy balances. Here we write the mass-balance relationships that must hold for all problems of this particular type:

$$V_1 + V_2 = V_3 \quad \text{(total liters must add up)} \tag{1.1}$$

$$f_1 V_1 + f_2 V_2 = f_3 V_3 \quad \text{(total alcohol must add up)} \tag{1.2}$$

- Next solve the general case. One way to do that is to set $V_2 = V_3 - V_1$ (by rearranging the first equation) and substitute to obtain

$$f_1 V_1 + f_2 (V_3 - V_1) = f_3 V_3.$$

Solve this for V_1, as follows:

$$f_1 V_1 + f_2 V_3 - f_2 V_1 = f_3 V_3$$

$$f_1 V_1 - f_2 V_1 = f_3 V_3 - f_2 V_3$$

$$(f_1 - f_2) V_1 = (f_3 - f_2) V_3$$

$$V_1 = \frac{f_3 - f_2}{f_1 - f_2} V_3 \quad \text{and} \quad V_2 = V_3 - V_1 \tag{1.3}$$

- Equation 1.3 is the general solution. There are computer tools, like MATLAB®, Maple®, Mathematica®, and Octave, that can do part of the work when we have such software available. This book will provide examples of using the first of those, but the others have similar capabilities. For useful general information on using MATLAB, see Hanselman and Littlefield (2001), and Higham and Higham (2000).

Even if such tools are available, however, we still must come up with the original relationships (Eqns. 1.1 and 1.2) by the logic of mass balances, and it is often useful to solve simple problems without cranking up the computer. It is also important to be able to check computer solutions "by hand."

For the present problem, to look into MATLAB a bit, if we entered the lines[1]

```
% Define some symbolic variables
syms V1 V2 V3 f1 f2 f3
soln=solve('V1+V2=V3','f1*V1+f2*V2=f3*V3',V1,V2)
V1=soln.V1
V2=soln.V2
```

MATLAB would return

```
V1=-V3*(-f3+f2)/(f1-f2)
V2=V3*(f1-f3)/(f1-f2)
```

which, with a little fiddling, can be put in a form identical to the solution found above (Eqn. 1.3). MATLAB would format the solutions differently if we entered

```
pretty(V1)
pretty(V2)
```

The result would be

```
  V3 (-f3 + f2)            V3 (f1 - f3)
- -------------    and     ------------
     f1 - f2                  f1 - f2
```

• Now is the time to put in the numbers for the particular case[2].

---

[1] Any material following a % sign in commands sent to MATLAB is treated as a comment, and ignored.

[2] In this text, a numeral with a bar over it, like the "$\bar{3}$" in Eqn. 1.4, indicates that the number under the bar is to be repeated *ad infinitum*. Thus, $0.5\bar{3}$ denotes the repeating decimal number $0.533333\ldots$; similarly, $0.6\overline{17}$ would represent the quantity $0.61717171717\ldots$.

$$V_1 = \left[\frac{(2/3) - 0.4}{0.9 - 0.4} \right] 1 = \frac{0.6\bar{6} - 0.4}{0.5} = 2(0.2\bar{6}) = 0.5\bar{3} \text{ liters} \quad (1.4)$$

$$V_2 = 1 - 0.5\bar{3} = 0.4\bar{6} \text{ liters}$$

This is the particular solution for the numerical values specified in this instance. It's a good idea to work most problems in this way—symbolically first, then substituting numbers at the end—because it produces general answers that can be reused, and that provide insight into the structure of the problem and its solution. Some people find it difficult to work in this way; if that is true for you, you may find it helpful to do an example set of numerical calculations before generalizing to the symbolic version.

For tough story problems, it often helps to use some of the following aids:

- List the units of all quantities involved. When a variable seems vague, this can help clarify what it is and what it means.

- Draw sketches. Try to represent the general nature of the solution with rough curves.

- Try a special case, e.g., with numbers instead of symbols, and then generalize to symbols.

- Solve a simpler problem by omitting complicating factors. Then, if you can solve the simple problem, add back the omitted factors, one at a time.

- For really hard problems, trial and error may help. Keep guessing at solutions and testing whether they meet all the conditions. Watching the patterns that develop for different guesses may help you to see what the general relationship is.

1.2   Pre-Calculus Math Review

If you wish to review basic pre-calculus math, have a look at Appendix A. In particular, if you are not comfortable working with logarithms[3], please review them there. I have made some arbitrary decisions about

---

[3]In this book, "log" will refer to the natural (base e) log; if base-10 logs are needed, they will be denoted by "$\log_{10}$." For more on this, see p. 288.

what material to put there and what to retain in this chapter, so don't be surprised to see material here that you may consider review.

1.3 Trigonometry.

Although the trigonometric functions ($\sin\theta$, etc.) are motivated by, and often defined in terms of, angles and sides of triangles, they have many uses in applied math that are independent of geometrical interpretations. It is often useful to think of them as periodic (repeating) functions of some arbitrary variable x. In applications, the independent variable is often time.

Note that although many people are accustomed to working with trigonometric functions with angles measured in degrees (360 degrees in a full circle), radians (2π in a full circle) are a more natural unit in mathematics. Unless stated otherwise, all angles used in this book will be expressed in radians. You should adopt this convention too. This means that you should figure out how to set the radian mode for trigonometric functions in your calculator.

The following exercises illustrate the idea of using sines and cosines to model periodic relationships.

1. Sketch these six functions and label both axes: a. $\sin x$; b. $\cos x$; c. $3\sin x$; d. $2\cos x$; e. $\sin(x + 5)$; f. $\cos(x - \pi/8)$.

2. How does the value of $y = a + b\cos[c(x - d)]$ change as a, b, c, and d change? A rough sketch of this function will help you to answer this question.

3. Consider the graph of $y = \sin(bt)$ shown in Fig. 1.1, p. 6, and determine the values asked about there. Note that the *period* of a sine or cosine function is the length of time required for the oscillation to complete one full cycle.

1.4 Units, Dimensions, and Conversion Factors

It may be that math becomes "applied math" when numbers have dimensions or units. *Dimensions* are concepts like time, mass, length, weight, etc. *Units* are specific cases of dimensions, like hour, gram,

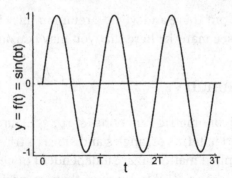

Figure 1.1: A plot of the function $y = \sin(bt)$. Here T is the *period* of $y = f(t)$; i.e., $T = 2\pi/b$. What values of b will be required to yield periods of a) 1 sec; b) 12 months; c) 24 hours; d) 1 year?

meter, lb_f, etc. As you know, you can *multiply* and *divide* quantities with different units:

$$3\,ft \times 7\,lb = 21 \text{ ft-lb}; \quad (70\,\text{mi})/(2\,\text{hr}) = 35 \text{ mi hr}^{-1},$$

but you can *add* and *subtract* terms only if they have the same units:

$$3 \text{ kg} + 2 \text{ lb}_m = TILT.$$

However, $3 \text{ kg} + 2 \text{ lb}_m \times \dfrac{453.6\,\text{g}}{\text{lb}_m} \times \dfrac{1\,\text{kg}}{1000\,\text{g}} = 3.91 \text{ kg}.$

You can use these rules to check formulas you derive; e.g., suppose you have derived the relationship $kT = c_p\rho V$. The variables here are

$$\left(\begin{array}{c}\text{thermal}\\\text{conductivity}\end{array}\right)\left(\text{temp}\right) = \left(\begin{array}{c}\text{specific}\\\text{heat}\end{array}\right)\left(\text{density}\right)\left(\text{volume}\right)$$

One set of units might be

$$\left[\frac{\text{mW}}{\text{cm}\cdot\text{deg}}\right]\left[\text{deg}\right] = \left[\frac{\text{mW}\cdot\text{sec}}{\text{g}\cdot\text{deg}}\right]\left[\frac{\text{g}}{\text{cm}^3}\right]\left[\text{cm}^3\right]. \text{ Does it check?}$$

Agreement of units is necessary (but not sufficient) for formulas to be correct.

Conversion Factors

Suppose you need to convert $2\,\text{cal}\,\text{cm}^{-2}\,\text{min}^{-1}$ into $\text{kW}\,\text{mi}^{-2}$. From tables, we find $1\,\text{cal min}^{-1} = 70\,\text{mW}$, $10^6\,\text{mW} = 1\,\text{kW}$, $2.54\,\text{cm} = 1\,\text{in}$, $12\,\text{in} = 1\,\text{ft}$, and $5280\,\text{ft} = 1\,\text{mi}$. Then

$$\frac{70\,\text{mW}}{1\,\frac{\text{cal}}{\text{min}}} = 70\,\text{mW min cal}^{-1} = 1 \text{ (dimensionless);}$$

$$\frac{1\,\text{kW}}{10^6\,\text{mW}} = 1 \text{ (dimensionless), etc.}$$

Thus,

$$2\frac{\text{cal}}{\text{cm}^2\,\text{min}} \cdot 70\frac{\text{mW min}}{\text{cal}} \cdot \frac{\text{kW}}{10^6\text{mW}} \cdot 2.54^2\frac{\text{cm}^2}{\text{in}^2} \cdot 12^2\frac{\text{in}^2}{\text{ft}^2} \cdot 5280^2\frac{\text{ft}^2}{\text{mi}^2}$$

$$\approx 3.63 \times 10^6 \text{ kw mi}^{-2}$$

You are probably familiar with unit conversions of this sort—they can be complicated, and require careful work.

In later chapters, we will need to work with the units of derivatives and integrals. Here are a few examples. If x = meters and t = seconds, the units of these quantities:

$$\frac{dx}{dt}, \quad \frac{d^2x}{dt^2}, \quad \left(\frac{dx}{dt}\right)^2, \quad \text{and } y = \int_{t_1}^{t} x^2\,dt$$

are m s^{-1}, m s^{-2}, $\text{m}^2\,\text{s}^{-2}$, and $\text{m}^2\,\text{s}$, respectively.

1.5 Ratios and Percentages

These types of quantities can be confusing, and in particular, denominators must be carefully specified. Examples:

- A child weighs 50 lbs. In the next year, she gains 20% in weight. The following year, she gains another 20%. What is her weight now 40% higher than at the start?

- A machine produces 8% defective items. You adjust it to produce only 6% defective items. Is that a 2% or 25% improvement?

- In 2003, an apple tree produced 200 lb of apples. In 2004, it produced 50% more than in 2003. In 2005, it produced 50% less than in 2004. Was the 2005 production the same as that in 2003?

1.6 Analysis (Finding Symbolic Solutions) versus Numerical Analysis

Example 1: Consider polynomials; i.e., equations of the form

$$y = a_0 + a_1x + a_2x^2 + a_3x^3 + a_4x^4 + \cdots + a_nx^n = P_n(x).$$

With $P_1(x)$, the polynomial is $y_0 = a_0 + a_1x$. Thus $x = (y_0 - a_0)/a_1$. (That's pretty simple.)

With $P_2(x)$, we have $y_0 = a_0 + a_1x + a_2x^2$. Then $x = ?$ To solve this analytically, you'll need to use the quadratic formula that you've probably seen before, but you'll likely have to substitute symbols to put the problem into the specific form you learned then.

The approximate energy balance for the roof of a car might be $70 = (5.4 \times 10^{-9})T^4 + 1.2(T - 295)$, in terms of roof temperature T [deg K]. The 70 is the solar and thermal radiation absorbed by the roof, the T^4 term is the reradiated energy, and the $T - 295$ term is the convective heat loss to the air. How do we solve for T? There is no general "quartic" formula comparable to the quadratic formula. This solution can only be obtained numerically, by methods we'll see in Chapter 10.

Example 2: Consider models involving sets of linear equations. How do you solve for the x values from each set of N equations?

- $N = 1$: $a_1x_1 = b_1$

- $N = 2$: $a_{11}x_1 + a_{12}x_2 = b_1$
  $a_{21}x_1 + a_{22}x_2 = b_2$

- $N = 3$: $a_{11}x_1 + a_{12}x_2 + a_{13}x_3 = b_1$
  $a_{21}x_1 + a_{22}x_2 + a_{23}x_3 = b_2$
  $a_{31}x_1 + a_{32}x_2 + a_{33}x_3 = b_3$

- $N = 4$: $a_{11}x_1 + a_{12}x_2 + a_{13}x_3 + a_{14}x_4 = b_1$
  $$\vdots$$
  $a_{41}x_1 + a_{42}x_2 + a_{43}x_3 + a_{44}x_4 = b_4$

With this class of relationships, for $N = 1$, x_1 is obviously b_1/a_1. For $N = 2$, you have no doubt learned to solve one of the equations for x_1 in terms of x_2 (say), to substitute that relationship into the other equation, and to solve the latter for x_2. Back substitution then

yields x_1. (You've seen an example of this in the mass-balance problem at the beginning of this chapter.) When $N = 3$ or higher, a similar
process would work in principle, but becomes unnecessarily tedious.
Instead, one would usually use a numerical method from Chapter 9
to find the x values when N is three or higher.

Example 3: Consider the *logistic equation* for population growth;
i.e., the differential equation (which we'll see in more detail in Chapter 4):

$$\frac{dN}{dt} = rN\left(\frac{K - N}{K}\right) \text{ with } N = N_0 \text{ at } t = 0. \tag{1.5}$$

Here r = growth rate ($= b - d$), N =population size, t = time, and
K = carrying capacity. This equation models population growth that
is nearly exponential (following $dN/dt = rN$) when N is small compared to K, because the factor in parentheses is nearly unity then. As
$N \to K$ and that factor approaches zero, the growth rate approaches
zero too.

The solution to Eqn. 1.5, an analytic solution, is

$$N(t) = \frac{KN_0}{N_0 + (K - N_0)e^{-rt}}. \tag{1.6}$$

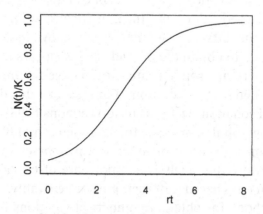

Figure 1.2: A generalized solution of the logistic population growth equation, with $N(t)$ represented as a fraction of the carrying capacity K, and
with the population starting out at $0.05K$.

Although specific numerical values are required to graph the solution as in Fig. 1.2, the analytic solution allows us to infer much of its
general behavior from the equation itself. For example,

- At $t = 0$, it is easy to check that $N(t) = N_0$.

- As $t \rightarrow \infty$, we see that $N(t) = K$.

- Because t enters the solution only through the product rt, we can see that if population A has twice the growth rate r compared to population B, but N_0 and K are the same for the two populations, then A will reach any given population size in half the time required by B.

Compare that model, for which an analytic solution is available, with a more complicated one, a modification of the *Lotka-Volterra* equations (p. 138), for the changes with time of two interacting populations. Here H might represent the biomass of some herbivore population, and C that of a carnivore population on some area of land.

$$\frac{dH}{dt} = r_H H \left(\frac{K_H - H}{K_H} \right) - aHC \text{ with } H = H_0 \text{ at } t = 0, \text{ and} \qquad (1.7)$$

$$\frac{dC}{dt} = r_C C \left(\frac{bH - C}{bH} \right) + cHC \text{ with } C = C_0 \text{ at } t = 0. \qquad (1.8)$$

In this model, the coefficients a, b, c, r_H, r_C, and K_H are often taken to be constants. Even so, no analytic solution of the form "$H(t) =$ explicit equation, $C(t) =$ explicit equation" is obtainable for this system of equations, but the numerical solution (for any particular set of constants) can be obtained, as shown in Fig. 1.3, p. 11. The method used to obtain the plotted values appears in Chapter 7.

Although there are sophisticated techniques for inferring some aspects of the behavior of solution of models like this directly from the differential equations (e.g., Mesterton-Gibbons 1995), the solution itself (including population sizes, for example) can only be obtained numerically, for a given set of numerical parameter values. These examples illustrate that analytic solutions are often more desirable than numerical ones because of their greater generality. However, we also need methods for obtaining numerical solutions for the many situations when analytic solutions are not available.

1.7 Notes on Significant Digits

In problems involving numerical calculations based on inexact data or on rough estimates, apparent conflicts sometimes arise between:

- The need to avoid round-off error during calculations, and

- The need to avoid an appearance of excessive certainty in the final values presented.

The following guidelines (developed in a discussion with my colleague, Ronald Hites) should help you in working with numbers and in reporting results.

1. *While performing calculations,* keep at least three more significant digits in all numbers than you will report in your final answer. In many calculations, like those involving matrices, it is best to keep all available digits—this helps reduce round-off error, which can be substantial, especially in a series of calculations.

2. When using a calculator, to the extent possible store intermediate calculations rather than writing them down and then re-entering them. This avoids transcription and re-entry errors, and retains full precision.

3. When reporting results, present one more significant digit than you had in your least precise input value. Be a little flexible here— note that 0.999 and 1.111, with 3 and 4 significant digits respectively, have nearly identical precision, while 0.999 and 9.999 differ in precision by an order of magnitude. As another example, the

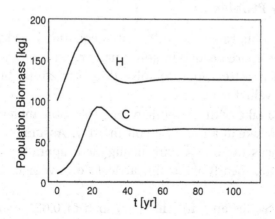

Figure 1.3: Results of a numerical solution of the Lotka-Volterra equations (Eqns. 1.7 and 1.8) with $r_H = 0.07$ yr^{-1}, $K_H = 2000$ kg, $a = 0.001$ kg^{-1} yr^{-1}, $r_C = 0.03$ yr^{-1}, $b = 0.1$ [-], and $c = 0.001$ kg^{-1} yr^{-1}.

sum of 0.333, 0.334, and 0.335 is 1.002. If the three original values are really known to three places, then valid information would be lost by rounding the sum to 1.00, yet that is what the "standard rule" that many learn says to do.

Also, statistical theory tells us that means of N numbers are $\sqrt{N}$ times more precise than the individual numbers being averaged. Thus, a mean of 100 numbers is known with an additional digit of precision compared with the original data. Even with fewer than 100 numbers, this tells us that the result of a calculation *can* have more precision than the individual items that go into it.

4. With measurements, three significant figures are usually about right for your final reported value. Use more only in special cases when you need, and can justify, more.

5. When using or presenting data with a ± error range, you should know (and state) what the error range represents. For example, is it a standard deviation, a confidence interval, or a maximum bias?

6. Measurements presented as 6.6 ± 0.4, say, seldom indicate a belief that the true value lies anywhere in the range from 6.2 to 7.0 with equal likelihood—instead the likelihood is concentrated near 6.6.

7. Above all, use common sense.

Preparatory Problem

This exercise illustrates why differential equations, which we'll take up in Chapter 4, are needed to solve many real problems in environmental science. Attempting to solve this example problem without that tool is a valuable exercise.

Consider a lake with a volume of 3×10^6 cubic meters. A stream flows into the lake at a rate of 2,500 m³/day. Assume that an outlet stream balances the inflow with negligible evaporation, so the lake volume remains nearly constant, at least on average, over several years.

The lake carries an initial mercury load of 0.025 mg/liter. Most of this has come in via the stream, which enters the lake with a mercury concentration of 0.3 mg/liter. To simplify matters, assume that all mercury is dissolved in the water and that none of it evaporates, drops to the bottom with sediments, is taken up by organisms, or is involved in chemical reactions.

Develop a method to compute a rough estimate of mercury concentration in the lake water after a period of five years, and then carry out the computations. Keep your method fairly simple. **Please do not use differential equations, even if you know how to use them.**

You should write out a) a list of any important assumptions and simplifications that you make to obtain an approximate solution. b) a brief description (about 1 page) of how your method operates to stimulate physical reality, c) a list defining all variables you use, and d) a statement of your results:

1. What approximations are involved in your scheme.

2. Whether your estimate is likely to be larger or smaller than the true value, and why.

3. What else you would need to know, if you wanted to improve your estimate. Work in symbols for as long as you can, and show your work.

1.8 Exercises

Algebra Story Problem Practice[4]

We'll start with some relatively straightforward story problems, as practice for the more complex ones in later chapters. In solving these problems, try to work in symbols as long as possible, so as to obtain general answers; then substitute numbers only at the end of each problem. However, if it works better for you, solve the problem with numbers first, and then derive the more general symbolic solution.

1. River City is 3/5 as far from Victor as Horner is from Victor. If River City is 80 miles from Victor, how far is it from Victor to Horner?

2. On a field trip, the professor for the course and the teaching assistant both drove the van. The prof drove 2.5 times the distance

---

[4]Answers to odd-numbered problems for all chapters may be found in Appendix B, pp. 292*ff*.

that the TA drove. How far did the TA drive if the prof drove 440 miles?

3. Two trucks haul materials to a landfill. The larger truck carries 3.8 tons each trip, or 2.6 times the weight carried by the smaller truck. What weight did the latter carry?

4. A experimental tank contained 8.2 kg of salt (NaCl), which made up 12% of the mass (weight) in the tank; the rest was water. What mass of water did the tank contain?

5. A rectangular field is 500 m by 900 m. A farmer plows around the perimeter of the field until 1/4 of the field's area is plowed. How wide a border has been plowed at that time?

6. Annie made a 100-mile round trip in 2.8 hours. Because of bad weather, she drove 7 mph slower on the return part of the trip than on the outgoing part. What was her speed each way?

7. Pump A alone can fill a storage tank in 2.5 hours less time than it would take pump B alone. If both pumps together can fill the tank in 6 hours, how long would it take each pump working alone?

8. Two volunteers for a public-interest group are able to insert 2200 letters into envelopes in 3.5 hours. How many letters could be dealt with in 2 hours by 11 volunteers working at the same rate?

9. On a map you are consulting, 40 miles corresponded to 0.8 inch. If two locations are 2.5 inches apart on the map, how far apart are they in reality?

10. A certain fraction has the property that if the same constant (*any* constant) is added to both its numerator and denominator, its value doesn't change. What is the original fraction?

11. A copper rod, part of an instrument, is exposed to varying temperatures. Its length L is almost a linear function of the temperature T, as long as T is less than 150°C. Find an equation for $L(T)$ using the following measurements: $(T = 15; L = 76.45$ cm) and $(T = 100; L = 76.56$ cm)

12. Suppose that one animal is 15% larger in every linear dimension than a second animal—the two then have the same shape (they are

geometrically similar). How do their surface areas, volumes and weights (under the assumption of constant specific gravity) differ?

13. Suppose an animal grows without a change in shape. If its volume increases by 40%, by what percentage does its surface area increase?

14. A population that grows exponentially (i.e., $N(t) = N_0 e^{rt}$) has a doubling time that remains constant. That is, the population would grow from 40 million to 80 million in the same amount of time it would take to grow from 5 million to 10 million. Thus, if N goes from N_0 to $2N_0$, then $2N_0/N_0 = 2 = e^{rD}$, where D must be the doubling time. Taking the natural log of both sides yields $\log 2 = rD$, or $D = (\log 2)/r$ (assuming r remains constant). (Be sure you understand why!)

Similarly, decay of radioactive substances often follows the model $M = M_0 e^{-rt}$. Scientists then refer to the half-life H; i.e., the length of time required for a radioactive mass to drop from any amount M to an amount $M/2$. Find an expression for H in terms of r.

15. Calculate, and plot (with speed S on the horizontal axis), the amount of time you would save by driving 100 miles at $(S + 10)$ miles per hour rather than at S miles per hour. Do this for $10 \le S \le 70$ mph.

16. If a certain length of wire is bent to form a square, will it enclose a larger or a smaller area than if it were bent to form a circle? How much larger or smaller?

17. For a study of temperature in birds' nests, you borrow an old instrument that uses thermistors to sense temperature. The manual for the instrument tells you that the resistance rises nearly linearly with the temperature, over a temperature range between 0–100° C, and that you should calibrate the system every six months or so to account for any aging of components. So, you obtain output readings of r_1 kΩ (kilohm) at T_1 °C and of r_2 kΩ at T_2 °C.

A. Find the equation (in terms of the symbols r_1, r_2, T_1, and T_2) that would give you temperature as a function of resistance in future measurements.

B. If $r_1 = 8.7$ kΩ and $r_2 = 12.5$ kΩ when $T_1 = 0$° C and $T_2 =$

40° C respectively, what would the sensor temperature be when the reading is 9.2 kΩ?

Hint: Of the various formulas for working with straight lines, the "two-point formula" is often the most efficient. If you know two specific points (x_1, y_1) and (x_2, y_2) on a straight line, and if (x, y) is any other point on the line, then the line can be defined by:

$$\frac{y - y_1}{x - x_1} = \frac{y_2 - y_1}{x_2 - x_1}.$$

This works because both sides of the equality are expressions for the slope of the line. You may use this if you like, but there are other ways to solve the problem.

Exercises involving ratios

The essence of mathematical modelling, and much of applied mathematics, is setting up word problems, converting them to mathematical equations, and then solving those. Here are three problems to solve. *Be careful*, because each involves ratios, and ratios are frequently tricky. Remember in the problems below to work in symbols for as long as you can.

18. B left Dandongadale 30 minutes after A, and travelled in the same direction. If A travelled 50 mph and B travelled 60 mph, how far had they gone when B overtook A?

19. A pollution control project has been running for 10 years. During its first 5 years the project had a benefit:cost ratio of 1.1. During the second 5 years, the ratio increased to 1.2. What is the overall benefit:cost ratio for the 10 year period? (The P.R. Department wants to know in 5 minutes!)

20. If you drive 50 miles at 40 mph, and 50 more miles at 50 mph, what is your average speed for the 100 mile trip?

Exercises with periodic functions

21. You are working with a group modelling forest growth, and you find that for your location, the amount of sunlight available (on cloudless days) varies roughly sinusoidally, with a maximum of

about 500 cal cm^{-2} day^{-1} on day 172 of the year and the minimum of about 160 cal cm^{-2} day^{-1} on day 354 of the year. To simplify matters, you may assume that all years have 364 days (to make it an even number). You decide to model this relationship using $Q = a + b \sin[c(t + d)]$, where Q is the radiation [cal cm^{-2} day^{-1}] and t is time [days] from the start of each year. (A rough sketch will doubtlessly be helpful.)

A. In terms of quantities given above, what are the values of a and b? What are the units of each?

B. What are the values of c and d? What are the units of each?

22. Mountain streams sometimes cause major problems for alpinists because of variability of water level during a day. In particular, some streams can be easily and safely crossed in early morning, but later in the day they become too high to cross because of increases in snowmelt as the air warms and solar radiation increases. The same phenomenon is of interest to hydrologists in mountain regions.

Suppose you want to model such stream flow as a sine wave with a period of one day, and with time measured in hours after midnight, using a 24-hour decimal clock (i.e., $t = 13.25$ at 1:15 p.m., $t = 23.5$ at 11:30 p.m., etc.). Suppose the minimum stream flow each day is about q_1 m^3 s^{-1} and the maximum is about q_2, with the maximum occurring at 3 p.m. ($t = 15$) each day. Determine the parameters (a, b, c, and d) that would make the function $q(t) = a + b \sin[c(t+d)]$ fit that situation. The $c(t+d)$ part should be in radians. (Radians are technically ratios of two lengths, and so have no units.)

In your final answer, a and b should be expressed in terms of q_1 and q_2, while c and d should have numerical values. If you prefer to work with numbers, you could start with numerical values $q_1 = 20$ and $q_2 = 100$, but your final answer should be based on general, symbolic values for these two quantities. You will almost certainly find it helpful to sketch the curve for which you are writing the equation.

1.9   Questions and Answers

These items, and those in similar sections in other chapters, are questions asked by students in some of my classes, and the answers I provided by e-mail.

1. You sometimes use the name "exp." What does that mean?

   • "exp" is one name for the exponential function; exp(anything) stands for the number e taken to the 'anything' power. That is, $\exp(x)$ is another name for e^x, and $\exp(\sin t)$ is another name for $e^{\sin t}$. It is convenient to use the "exp" form when you don't want to have to type an exponent as a superscript, especially when the exponent is a complicated expression. You'll see this a lot in books and reports. See p. 287.

2. Does the sine of 0 equal 1?

   • No. Actually, $\sin(0) = 0$. However, $\cos(0) = 1$.

3. Why is (m/s)/s, that is, meters per second per second, equal to $m/(s^2)$, that is, meters per second^2?

   • (m/s)/s is the same as (m/s) multiplied by the fraction (1/s). If you multiply two fractions (say 2/3 and 4/5) then the result is the products of the numerators (2*4) divided by the product of the denominators (3*5). For the (m/s)*(1/s), that becomes (m*1)/(s*s), which is $m/(s^2)$. What you might call "multiple divisions" like this can be confusing. That's why most scientific journals these days want units of quantities like milligrams per square meter per hour (a deposition rate of some quantity to a soil surface, say) to be printed as mg m^{-2} hr^{-1} rather than as mg/m^2/hr.

Chapter 2

Derivatives and Differentiation

2.1 What Is a Derivative?

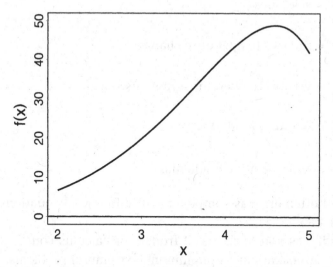

Figure 2.1: A generalized function, for use in illustrating the definition of the first derivative.

When asked "What is the derivative at a point x of the function $y = f(x)$ plotted in Fig. 2.1," students most often answer "the slope of the line above that x." That geometric interpretation is correct, but a more mathematical definition is

$$\frac{dy}{dx} = \lim_{h \to 0} \frac{f(x+h) - f(x)}{h}.$$

Copying the figure and sketching the line from, say, $f(4)$ to $f(4+h)$ with $h = 1$, then $h = 0.5$, and then $h = 0.1$ may help you to see the relationship between those two definitions.

2.2   Usefulness in Environmental Science

Derivatives arise in environmental science in two general ways. First, many derivatives are fundamentally important; e.g.,:

$$\frac{dN}{dt} = \text{rates of population change} \tag{2.1}$$

$$\frac{dm}{dt} = \text{mass flow rates} \tag{2.2}$$

$$\frac{dx}{dt} = \text{velocities} \tag{2.3}$$

$$\frac{dT}{dt} = \text{rates of temperature change} \tag{2.4}$$

$$\frac{dT}{dz} = \text{vertical temperature gradients (lapse rates)} \tag{2.5}$$

$$\frac{dP}{dz} = \text{pressure gradients} \tag{2.6}$$

$$\frac{dC}{dx} = \text{concentration gradients.} \tag{2.7}$$

These will often arise as components of differential equations, as in Chapters 4 and 5.

Secondly, as you likely recall from your calculus course, derivatives arise in maximum and minimum (extremum) problems. Recall that the derivative of a function is zero at every local or global maximum or minimum point, as in Fig. 2.2, p. 21. Let's consider an example max-min problem, after first recalling some basic relationships.

- First, for any function $y = x^p$, the derivative $dy/dx = px^{p-1}$.

- The derivative of the sum of several terms is the sum of their derivatives.

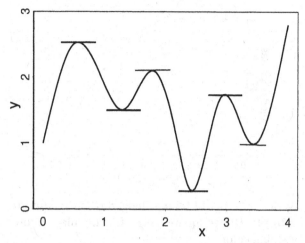

Figure 2.2: Graph of a function with six maxima and minima in the range from $x = 0$ to $x = 4$. What is the value of the derivative (dy/dx) at each of these extreme points?

- Recall the quadratic formula. If $f(x) = ax^2 + bx + c$, then $f(x) = 0$ when $x = (-b \pm \sqrt{b^2 - 4ac})/2a$.

Here's the example problem: An ornithologist studying the (fictional) black-booted albatross goes to twenty breeding colonies and measures breeding success as a function of how densely packed the breeding pairs are in the various colonies. She finds by polynomial regression that the relationship can be approximated by

$$F = A + BD + CD^2,   \tag{2.8}$$

where F is the average number of young fledged (successfully raised) per breeding pair, D is the density of breeding pairs in the colony (pairs m^{-2}), $A = 4$, $B = 2$, and $C = -2$.[1] This relationship is shown in Fig. 2.3, p. 22. Because total area and suitable locations for breeding of this species are limited, the researcher asks you to estimate the breeding-pair density that would produce the maximal number of young fledged per unit area of colony, assuming that Eqn. 2.8 is reasonably accurate.

If F is young pair^{-1} and D is pairs m^{-2}, then $S = FD$ gives the number of young per square meter (check the units). The density

---

[1]The units of these coefficients are not particularly useful to work with, but are whatever they need to be to make the equation come out right.

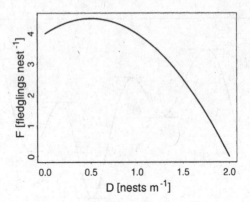

Figure 2.3: Average number of young fledged (F) in relation to number of breeding pairs (nests, D) per square meter, for the "black-booted albatross" at different breeding colonies.

D that maximizes $S = FD = AD + BD^2 + CD^3$ is the value for which $dS/dD = A + 2BD + 3CD^2 = 0$. Applying the quadratic formula to that equation yields $D = (-B \pm \sqrt{B^2 - 3AC})/(3C)$. Substituting numbers yields values of about 1.215 and -0.549 pairs m^{-2} for D, of which only the positive value makes sense. For that D, the success rate S is about 4.23 young m^{-2}.

General Tips for Solving Max-Min Problems

When you want to find the value of some variable x that causes another variable y to be a maximum or a minimum, the following steps may help:

1. Read the problem, and state in your own words what you know and what you are trying to find. With the albatross problem, we could start with a particular number of pairs per m^2, and then could use the equation for F to get the number of young per breeding pair. What we don't know is the number of pairs per m^2 that would make the areal success a maximum.

2. Draw a diagram, label it, and identify what is constant and what varies. Here we might guess at a curve of S plotted against D. Clearly the curve should start at $S = 0$ when the breeding-pair density is zero. A little thought would show that since the young per nest goes to zero when adults become too dense, S must be

zero for $D \geq 2$ as well. (See Fig. 2.3.) Thus, you can guess that the curve reaches some maximum for D between 0 and 2 pairs per m^2.

3. Always pay careful attention to units! Units are often the key to a quick solution. Here S [young m^{-2}] must equal D [pairs m^{-2}] times F [young pair^{-1}].

4. a. Try to write an equation of the form $y = f(x)$, where y is the quantity to be maximized or minimized, and x is the quantity you can control. Here we would want S as a function of D, and the units tell us directly that $S = DF$.

   b. If there is more than one quantity that you can control, such as x and z, then write an equation of the form $y = g(x, z)$. Then search for a relationship between x and z that allows you to elim-inate z, and to convert $y = g(x, z)$ to $y = f(x)$. (This is not needed for the present example.)

5. Solve for the value of x that makes $df/dx = 0$. Then determine whether y is a maximum or a minimum for that x value.

6. State your conclusions in words.

Straightforward Nature of Differentiation

Differentiation can always be accomplished analytically (i.e., in terms of symbols), by applying various definite rules. First, of course, one needs to know certain basic derivatives, such as those of x^p, e^x, e^{bx}, $\log x$, $\log bx$, $\sin bx$, and $\cos bx$. (Here b and p are constants, and x is a variable.) Try writing down the derivatives of those functions; then check your answers using any calculus text, or a table like that in the *Handbook of Chemistry and Physics* (Lide 2005). Be sure to note any exceptions to general rules.

   Those specific derivatives follow from the definition of the deriva-tive; i.e., from[2]

$$\frac{df(x)}{dx} \stackrel{\text{def}}{=} \lim_{h \to 0} \frac{f(x + h) - f(x)}{h}.$$

---

[2]The symbol " $\stackrel{\text{def}}{=}$ " should be read as "is equal to by definition." Distinguishing equality by definition from equality that follows from a series of mathematical steps can often aid understanding.

For example, when you studied introductory calculus, you probably carried out a series of calculations something like these, which show that the derivative of x^3 is $3x^2$:

$$\frac{dx^3}{dx} = \lim_{h \to 0} \frac{(x+h)^3 - x^3}{h}$$

$$= \lim_{h \to 0} \frac{x^3 + 3x^2h + 3xh^2 + h^3 - x^3}{h}$$

$$= \lim_{h \to 0} 3x^2 + 3xh + h^2 \ = \ 3x^2$$

Similar though often more complicated calculations lead to other derivatives.

Environmental scientists using math as a tool can work most efficiently if they memorize and know how to use these most common derivatives, along with the rules that follow soon for differentiating combinations of functions. However, computer software that can perform symbolic (analytic) calculations is becoming more readily available, and is worth learning too. As a simple example for now, here's how to obtain the analytic derivative of $f(x) = \sin(bx)$ using MATLAB®[3].

To differentiate that function, we enter the following lines (the parts to the left of the dots, that is) in the MATLAB command window:

syms b x ..... Treat b and x as symbols rather than numeric values.
f=sin(b*x) .... Define the function.
diff(f) .... Perform the differentiation.

After you enter the third line, MATLAB returns the result in the form ans = cos(b*x)*b.

Interestingly, we didn't have to tell the program that x was the variable and b a constant. Here's the reason, taken from the MATLAB help system: "The default symbolic variable in a symbolic expression is the letter that is closest to 'x' alphabetically. If there are two equally close, the letter later in the alphabet is chosen." I recommend trying your hand at using MATLAB (or a similar program) to obtain the other derivatives listed above and below.

Important rules that aid in differentiation are the sum, product, and quotient rules:

---

[3] This assumes availability of a version of MATLAB that includes the "Symbolic Math Toolbox."

$$\frac{d(u + w)}{dx} = \frac{du}{dx} + \frac{dw}{dx}$$

$$\frac{d(uw)}{dx} = w\frac{du}{dx} + u\frac{dw}{dx} \tag{2.9}$$

$$\frac{d(u/w)}{dx} = \left(w\frac{du}{dx} - u\frac{dw}{dx}\right)/w^2$$

One rule we'll use fairly often is the *chain rule*, which helps us deal with functions of functions. For example, if $y = f(x)$ and $z = g(y)$, then the dependence of z on x can be determined from $z = g[f(x)]$. If we want to know how rapidly z changes with changes in x, we could obtain dz/dx from the chain rule,

$$\frac{dz}{dx} = \frac{dz(y)}{dy}\frac{dy(x)}{dx} = \frac{dg(y)}{dy}\frac{df(x)}{dx}$$

For example, if $z = y^3$ and $y = e^x$, then $dz/dx = 3e^{3x}$. (Check this out for yourself.)

To see how the chain rule might arise in practice, suppose the turbidity T of the water in a stream is a function $T = f(C)$, where C is the concentration of clay particles in the water, and that C in turn is a function $C = g(V)$ of the stream velocity V. Now, suppose you wanted to know how much turbidity T increases for a unit increase in stream velocity V. That is, you want dT/dV. However, the functions we know are f and g. To get dT/dV, we use the chain rule, here $dT/dV = (dT/dC) \times (dC/dV)$. Because $T = f(C)$ and $C = g(V)$, we can also write $dT/dV = (df/dC) \times (dg/dV)$—this is just another way of saying the same thing.

Often we need to combine rules. For example, let's differentiate

$$y(x) = \frac{x^3}{1 + e^{3x}}$$

with a series of elemental steps to illustrate the process. Experienced mathematicians might perform many of these steps "in their heads," but here I illustrate the process in a way that even a novice can use safely:

Let $u \overset{\text{def}}{=} x^3$, from which $du/dx = 3x^2$

Let $v \overset{\text{def}}{=} 1 + e^{3x}$, $s \overset{\text{def}}{=} 1$, and t $\overset{\text{def}}{=} e^{3x}$ so v = s + t. Then

$$\frac{ds}{dx} = 0, \text{ so } \frac{dt}{dx} = 3e^{3x}, \text{ and}$$

$$\frac{dv}{dx} = \frac{ds}{dx} + \frac{dt}{dx} = 3e^{3x}.$$

Finally,

$$\frac{dy}{dx} = \frac{v\dfrac{du}{dx} - u\dfrac{dv}{dx}}{v^2} = \frac{(1 + e^{3x})(3x^2) - (x^3)(3e^{3x})}{(1 + e^{3x})^2}.$$

The point is, you can combine the various rules to differentiate *any* function analytically. In contrast, there are many functions that cannot be *integrated* analytically, as we shall see in Chapter 3.

For another example, it is possible to differentiate

$$f(x) \stackrel{\text{def}}{=} \frac{\sqrt{\cosh\{\sin[\log(x + 1)]\}}}{(\sqrt{x} + 1)^2}.$$

In fact, differentiation of this function is straightforward (although by no means simple) in the sense that it can proceed by a series of definite, sequential steps. We will skip the details, however, and note that there's an easier way to obtain this derivative if we have access to computer software that can perform *symbolic* math. (This functionality is *not* available in current spreadsheets and similar programs, but is provided by Maple, Mathematica, MATLAB, and some other programs.) To obtain the derivative using MATLAB, we could enter[4]

```
syms x...................................................Make x symbolic
f=sqrt(cosh(sin(log(x+1))))/((sqrt(x)+1)^2) ........Define f
diff(f) .......................................... Find the derivative
```
    The result would be returned as

```
ans =
1/2/cosh(sin(log(x+1)))^(1/2)/(x^(1/2)+1)^2*
sinh(sin(log(x+1)))*cos(log(x+1))/(x+1)-
cosh(sin(log(x+1)))^(1/2)/(x^(1/2)+1)^3/x^(1/2)
```

Entering pretty(ans) "prettyprints" the answer in a somewhat more readable form as

---

[4]In MATLAB, "log" refers to the *natural*, not common, logarithm.

```
           sinh(sin(log(x + 1))) cos(log(x + 1))
 1/2  ----------------------------------------------  -
                    1/2   1/2      2
     cosh(sin(log(x + 1)))    (x    + 1) (x + 1)


                                              1/2
                          cosh(sin(log(x + 1)))
                          ------------------------
                            1/2      3  1/2
                           (x    + 1)  x
```

The probability of correctly differentiating a function this complex by hand on the first try is pretty small for most of us, and even if we use MATLAB, we shouldn't trust the result absolutely. Symbolic programs sometimes make mistakes, and anyway, we might have mistyped something. Besides, we could easily have made a mistake copying down the answer, so how could we check this result? **This latter question is one you should consider every time you obtain any mathematical result—how can you demonstrate, to yourself and others, that a result you have just obtained is correct?** That is one of the subjects of § 2.4, but first we'll take up some background material that we'll need there and later in the book.

2.3   Taylor Series; a Basis for Numerical Analysis

We now take a brief side trip from derivatives *per se* to consider a mathematical relationship that underlies much mathematical analysis, and (of interest to us) forms the basis for many methods of numerical analysis. We consider only the basic ideas; for more information on the present topic, see the sections on "Taylor polynomials", "infinite series", and "Taylor series" in some calculus text.

In applied math, we are often interested in "nice" functions, for which $f(x)$ and all needed derivatives exist and are continuous over the range of interest. There may be a few singularities where the series is not valid, as in the function

$$f(x) = \frac{1}{1 - x^2}$$

at $x = \pm 1$, but many of the functions we work with are defined for all positive x, at least.

Nice functions can be expanded in *Taylor Series*, which have the form:

$$f(x) = f(a) + \frac{f'(a)}{1!}(x - a) + \frac{f''(a)}{2!(x - a)^2} +$$

$$\frac{f'''(a)}{3!}(x - a)^3 + \ldots = \sum_{n=0}^{\infty} \frac{f^{(n)}(a)}{n!}(x - a)^n. \quad (2.10)$$

Note (because $0! = 1$) that the first term can be written as

$$f(a) = \frac{f(a)}{0!}(x - a)^0,$$

so it fits into the same pattern as all the other terms.

In this expression, the symbol f'' denotes the second derivative of f, f''' is the third derivative, and $f^{(n)}$ is the nth derivative. Also, recall that $N! = 1 \times 2 \times \ldots \times N$ for all integer $N \geq 1$. Remember too that $0! = 1! = 1$ by definition—this may seem odd, but it turns out to be convenient in many situations. It is also consistent with the *gamma function* of higher mathematics, which is related to factorials by the relation $\Gamma(x + 1) = x!$ when x is an integer ≥ 0. (The gamma function is more general than factorials, being defined even for non-integer values of x. It is part of many important statistical distributions.)

In words, Eqn. 2.10 states the remarkable fact that the value of a function (left-hand side) can be determined everywhere if you know its value and the value of all its derivatives at a single point $x = a$ (right-hand side). The equality holds within some "radius of convergence" R; i.e., it is valid for $|x-a| < R$. We say that $f(x)$ is "expanded about a."

In the particular case when $a = 0$, the series is called a *Maclaurin series*, and it then takes the form:

$$f(x) = f(0) + xf'(0) + \frac{x^2}{2}f''(0) + \frac{x^3}{6}f'''(0) + \ldots.$$

Some important Maclaurin series, ones that are sometimes considered in higher math to be the definitions of the functions they represent, are:

$$e^x = \exp(x) = 1 + x + \frac{x^2}{2!} + \frac{x^3}{3!} + \frac{x^4}{4!} + \ldots + \frac{x^n}{n!} + \ldots \quad (2.11)$$

$$\cos x = 1 - \frac{x^2}{2!} + \frac{x^4}{4!} - \frac{x^6}{6!} + \ldots$$

$$\sin x = x - \frac{x^3}{3!} + \frac{x^5}{5!} - \frac{x^7}{7!} + \ldots$$

Note how these three series involve the same kinds of terms, but in different combinations and with different signs. Exercise 14, p. 41, demonstrates some interesting implications of these series.

Taylor Series Example

We won't often use Taylor series directly and explicitly in this book, but familiarity with them is useful because they provide important background for many of the analyses we will perform. To see how the series work, we consider a simple function here to allow comparing our results with easily calculated values. The Taylor series for $f(x) = \sqrt{x}$ (or $x^{1/2}$) expanded about $a = 4$ is

$$f(x) = \sqrt{x} = f(4) + f'(4)(x-4) + \frac{f''(4)}{2!}(x-4)^2 + \frac{f'''(4)}{3!}(x-4)^3 + \ldots$$

We can simplify this, but must do so carefully. To get $f'(4)$, first differentiate $f(x)$ symbolically, and *then* substitute 4 for x. That is, you must differentiate $\sqrt{x}$ for the variable x, not for the constant value, 4. The same principle holds for the higher derivatives. In symbols, then[5]:

$$f(x) = x^{1/2}; \quad f'(x) = \frac{df}{dx} = \frac{dx^{1/2}}{dx} = \frac{1}{2}x^{-1/2} = \frac{1}{2}\frac{1}{\sqrt{x}}$$

$$f''(x) = -\frac{1}{4}x^{-3/2}; \quad f'''(x) = \frac{3}{8}x^{-5/2};$$

$$f^{(4)}(x) = -\frac{15}{16}x^{-7/2}; \text{ etc.}$$

Now substitute $x = 4$ (because $a = 4$):

$$f(4) = 2; \quad f'(4) = \frac{1}{2\sqrt{4}} = \frac{1}{(2)(2)} = \frac{1}{4}; \quad f''(4) = -\frac{1}{(4)(8)} = -\frac{1}{32}$$

$$f'''(4) = \frac{3}{(8)(32)} = \frac{3}{256}; \quad f^{(4)}(4) = -\frac{15}{(16)(128)} = -\frac{15}{2048}$$

---

[5] Recall that the derivative of x^p is px^{p-1}.

Putting this all together yields the Taylor series for $f(x) = \sqrt{x}$ expanded around $a = 4$:

$$f(x) = 2 + \frac{1}{4}(x-4) - \frac{1}{(2)(32)}(x-4)^2 + \frac{3}{(6)(256)}(x-4)^3 - \ldots, \text{ or}$$

$$f(x) = \sqrt{x} = 2 + (x-4)/4 - (x-4)^2/64 + 3(x-4)^3/1536 - \ldots \quad (2.12)$$

for all $x \geq 0$.

To use this series, we know that $f(4) = \sqrt{4} = 2$. Thus to get $f(4.1)$ (i.e., $\sqrt{4.1}$), calculate

$$f(4.1) = \sqrt{4.1} = 2 + \frac{(4.1-4)}{4} - \frac{(4.1-4)^2}{64} + \frac{3(4.1-4)^3}{1536} - \ldots$$

$$= \sqrt{4.1} = 2 + \frac{0.1}{4} - \frac{0.01}{64} + \frac{0.001(3)}{1536} - \ldots, \text{ or}$$

$$\sqrt{4.1} = 2 + 0.025 - 0.00015625 + 0.000001953 - \ldots$$

$$\approx 2.024845703.$$

Compare that with the direct result $\sqrt{4.1} = 2.024845673$ from a calculator.

Now, for various x values try different polynomial orders (i.e., terminate the infinite series at earlier and earlier terms):

$$(n = 3): \quad f(x) \approx 2 + \frac{1}{4}(x-4) - \frac{1}{64}(x-4)^2 + \frac{3}{1536}(x-4)^3$$

$$(n = 2): \quad f(x) \approx 2 + \frac{1}{4}(x-4) - \frac{1}{64}(x-4)^2$$

$$(n = 1): \quad f(x) \approx 2 + \frac{1}{4}(x-4)$$

$$(n = 0): \quad f(x) \approx 2$$

Table 2.1, p. 31, shows the relative error; i.e.,

$$\text{relative error} \overset{\text{def}}{=} \frac{\text{approx } f(x) - \text{true } f(x)}{\text{true } f(x)} \overset{\text{def}}{=} \frac{\text{absolute error}}{\text{true value}}$$

that results from using different levels of approximation with various values of $x - a$. In the present case, the relative error is

$$\text{RE} = \frac{(\text{Taylor series for } \sqrt{x}) - \sqrt{x}}{\sqrt{x}}.$$

Table 2.1: Relative errors resulting from approximating $f(x) = \sqrt{x}$ for different values of x (columns), using the Taylor series of Eqn. 2.12 with different numbers of terms (rows). The square root function is expanded about $a = 4$. $n = 4$ corresponds with truncating the series at the cubic term, etc.

$x - a$:	0	0.1	0.4	4	36
n	$x = 4$	$x = 4.1$	$x = 4.4$	$x = 8$	$x = 40$
4	0	1.5×10^{-8}	3.5×10^{-6}	0.016	12.0
3	0	-9.5×10^{-7}	-5.6×10^{-5}	-0.028	-2.5
2	0	7.6×10^{-5}	0.0011	0.061	0.74
1	0	-0.012	-0.047	-0.29	-0.68

Note that the relative errors decrease in size as the number of terms increases except when $x - a = 36$. This occurs because when x is far from a, $(x-a)^k$ in the numerator of terms increases faster than $k!$ in the denominator, until eventually k becomes large enough. This table illustrates (though it can't be considered a proof) that when using a few terms of a Taylor Series (or some other polynomial) to approximate a function:

- evaluating the function at a point close to the expansion point ($x \approx a$) reduces error, and

- for x values close enough to a, more terms tend to yield greater accuracy. On the other hand, if $|x - a|$ is large, then adding a few terms can increase the error. Many more terms would be required before the series would begin to converge for $x = 40$.

The take-home point is that approximating functions over small intervals is generally desirable.

Eqn. (2.12) can be rewritten as $f(x) =$

$+2$	constant term
$-1 \quad +0.25x$	$(x - 4)/4$
$-0.25 \quad +0.125x \quad -0.015625x^2$	$(x - 4)^2/64$
$-0.125 + 0.09375x - 0.023438x^2 + 0.00195312x^3$	$3(x - 4)^3/1536$
$+ \ldots$	

This is equivalent to $f(x) \approx 0.625 + 0.46875x - 0.039062x^2 + 0.0019531x^3$, which illustrates that a Taylor series truncated after $n + 1$ terms is an nth order polynomial.

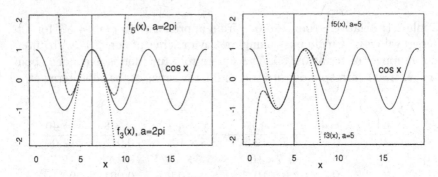

Figure 2.4: Two Taylor-series approximations to the cosine function, with $a = 2\pi$ (left) and $a = 5$ (right). The vertical line marks the value of a in each case. Note that the approximation based on the first six terms (f_5, Eqn. 2.14) approximates the function well over a wider range than the one based on only the first four terms (f_3, Eqn. 2.13).

Any "nice" function (one that is continuous, with all necessary derivatives also defined and continuous) can be written as an infinite polynomial of the general form $f(x) = a + bx + cx^2 + dx^3 + \ldots$. We will make frequent use of low-order polynomials to approximate more complicated functions. The theory of Taylor series, which we have barely touched on, provides the motivation.

As a further example, consider approximating the cosine function with terms through the third and fifth order of its Taylor series. That is, take $f(x) \approx \cos x$, where:

$$f_3(x) \overset{\text{def}}{=} \cos a - \frac{\sin a}{1!}(x-a) - \frac{\cos a}{2!}(x-a)^2 + \frac{\sin a}{3!}(x-a)^3, \quad (2.13)$$

and, adding two more terms,

$$f_5(x) \overset{\text{def}}{=} f_3(x) + \frac{\cos a}{4!}(x-a)^4 - \frac{\sin a}{5!}(x-a)^5. \quad (2.14)$$

The plots in Fig. 2.4 compare these approximations with the actual cosine function, for $a = 2\pi$ and $a = 5$, respectively.

2.4 Numerical Differentiation

In Chapter 3, we look at analytic integration, but then derive methods for calculating numerical values for definite integrals. Because not all functions can be integrated analytically, such numerical methods are essential tools in applied math.

Even though, as just discussed, all practical functions can be differentiated analytically (except at points of discontinuity), there are several reasons why we sometimes want to differentiate numerically as well:

- We may need the derivative of a function that we know only as a table of values of the form $[x, f(x)]$. For example, this situation might arise if we had a table of daily measurements of the volume of water in a reservoir.

- If a function is very messy, and we need its derivative at only one point, numerical differentiation may be the easiest way to obtain it. For example, the temperature distribution along the length of a cooling fin in a heat exchanger might be of the form

$$T(x) = T_a$$
$$+ (T_0 - T_a)\frac{\cosh m(L - x) + (h/mk)\sinh m(L - x)}{\cosh mL + (h/mk)\sinh mL}.$$

  If we needed dT/dx only at $x = 0$ as a step in calculating the total rate of heat loss from the fin, numerical differentiation might produce an acceptable answer most quickly.

- Numerical differentiation can form the basis for numerical methods for solving differential equations. We take up this topic later.

- Numerical derivatives can be very useful for checking whether an analytic derivative is correct or not. An example will follow shortly after we see how to do it.

To carry out numerical differentiation, first consider how we might approximate the slope of a function $y = f(x)$ at a particular point x_0. One method, called the *forward difference* approximation, is illustrated with Fig. 2.5 (left), p. 34. With this method, the derivative at $x = x_0$ is approximately

$$\frac{dy}{dx} \approx \frac{y(x_0 + h) - y(x_0)}{h}.$$

The forward difference method would be exact only for a linear function, $y = a + bx$.

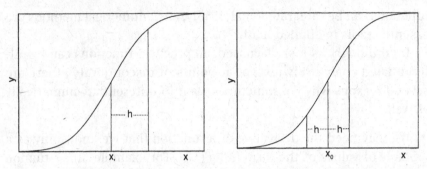

Figure 2.5: Graphs of a generalized function illustrating the *forward difference* numerical derivative (left) and the *central difference* numerical derivative (right).

A better method is the *central difference* scheme as defined with the aid of Fig. 2.5 (right). That approximation (for the derivative at $x = x_0$) is

$$\frac{dy}{dx} \approx \frac{y(x_0 + h) - y(x_0 - h)}{2h}.$$ (2.15)

The central difference formula is exact for a quadratic. To show that, let $y = f(x) = a + bx + cx^2$. Then

$$y' \; ? =? \; \frac{f(x + h) - f(x - h)}{2h}$$

$$= \frac{[a + b(x + h) + c(x + h)^2] - [a + b(x - h) + c(x - h)^2]}{2h}$$

$$= \frac{a + bx + bh + cx^2 + 2cxh + ch^2}{2h}$$

$$+ \frac{-a - bx + bh - cx^2 + 2cxh - ch^2}{2h}$$

$$= \frac{2bh + 4cxh}{2h} = b + 2cx.$$

But analytically[6],

$$\frac{dy}{dx} = \frac{d}{dx}(a + bx + cx^2) = b + 2cx. \quad \text{QED}$$

---

[6]QED, often found at the end of mathematics proofs, is an abbreviation for the Latin phrase *"quod erat demonstrandum"*, meaning "which was to be demonstrated."

For an example use of central differences, suppose you can't remember whether the derivative of $\cos x$ is $\sin x$ or $-\sin x$. We know from its graph that the sine function is positive just above zero, so what is $d(\cos x)/dx$ at $x = 0.1$? Using a hand calculator, our formula for the numerical derivative, and $h = 0.001$, we find

$$\frac{d(\cos x)}{dx} \approx \frac{f(x+h) - f(x-h)}{2h}$$

$$= \frac{\cos 0.101 - \cos 0.099}{0.002} = -0.09983.$$

Thus, we can conclude that our required analytic derivative is $-\sin x$.

Numerical derivatives suffer from and make good examples for demonstrating round-off error, which affects most numerical calculations. Because the central difference formula is correct only for quadratics, it is tempting to keep h very small when we apply the formula to higher-order functions. But the smaller we make h, the more alike will be the two terms in the numerator. When their difference is computed to a finite number of digits (as is true in all calculators and computers) a great deal of precision can be lost.

Let us demonstrate this by using our formula to estimate the derivative of $\sin x$ at $x = 1$. We know the result analytically; i.e., at $x = 1$, $d \sin x/dx = \cos 1 = 0.540302306$, and we can compare our estimates with that. If we choose $h = 10^{-5}$ on a machine with 12 digits, we obtain

$$\frac{d \sin(x)}{dx}\bigg]_{x=1} \approx \frac{\sin 1.00001 - \sin 0.99999}{0.00002}$$

$$= \frac{0.8414\hat{}76387773 - 0.8414\hat{}65581743}{0.0000200000000000}$$

$$= \frac{0.000010806030}{0.0000200000000000} = 0.54030(1500).$$

(The caret marks the point where the two terms begin to differ, and digits in parentheses are nonsense digits lost to round-off error.)

If we carry out similar calculations for various values of h, we find the results in Table 2.2, p. 36, which illustrate a compromise—as we go to smaller values of h, the round-off error increases, and the so-called *truncation error* (caused by truncating the Taylor series at the quadratic term) decreases because we stay ever closer to the point of expansion. One of the exercises at the end of the chapter will help you to choose a reasonable h to use with your own calculator.

Table 2.2: Effects of varying h on truncation error and round-off error, with the divided difference approximation to a derivative.

h	$d(\sin x)/dx$ (numerical)	truncation error	round-off error
1	0.4(54648713)	larger	smaller
10^{-1}	0.53(9402252)		
10^{-2}	0.5402(93300)		
10^{-3}	0.540302(220)		
10^{-4}	0.5403023(30)		
10^{-5}	0.54030(1500)		
10^{-6}	0.5402(93500)		
10^{-7}	0.5403(1)		
10^{-8}	0.53(98)		
10^{-9}	0.5(34)	smaller	larger

2.5   Checking Analytic Derivatives

As we will see later (e.g., p. 173), the function $\sinh x \stackrel{\text{def}}{=} (e^x - e^{-x})/2$, the *hyperbolic sine* of x, arises as a solution of second-order differential equations that describe processes like diffusion of substances and transfer of heat in the environment. If we differentiate this function analytically, we obtain

$$\frac{d \sinh x}{dx} = \frac{d}{dx}\left(\frac{e^x - e^{-x}}{2}\right) = \frac{e^x + e^{-x}}{2} \stackrel{\text{def}}{=} \cosh x,$$

the hyperbolic cosine function[7]. To check our answer, we could (a) calculate the numerical derivative at, say, $x = 1$, with $h = 0.001$, (b) calculate the numerical value of the analytic derivative at $x = 1$, and (c) compare the two. The numerical derivative of $\sinh x$ at 1 is:

$$\frac{\sinh 1.001 - \sinh 0.999}{0.002} = 1.5430809.$$

Our analytic derivative becomes

$$\frac{e^1 + e^{-1}}{2} = 1.5430806,$$

which checks pretty well!

---

[7]It happens that $\sinh' x = +\cosh x$ and $\cosh' x = +\sinh x$, which is similar to the parallel relationship with the circular sin and cos, but with no change of signs.

2.6 Exercises

Problem Set on Derivatives

1. Determine the derivative of the following function analytically, then find its numerical value when $a = 2\pi/12$, $b = 7$, $c = 0.01$, $g = 2$, $k = 3$ and $t = 3$.

$$f(t) = \frac{e^{-ct} \cos[a(t + b)]}{gt^2 + k}$$

2. Check your result to Exercise 1 using numerical differentiation, with the central difference of Eqn. 2.15, p. 34.

3. Calculate the actual (analytical) value of $d(\cos x)/dx$ at $x = 2$ and then check the result using

$$\frac{df}{dx} \approx \frac{f(x + h) - f(x - h)}{2h},$$

with $h = 1$, 10^{-2}, 10^{-4}, 10^{-6}, and 10^{-8}. Your goal is to determine a good h to use in your own calculator for similar calculations. When you locate the best value from those above, say 10^{-N}, try $10^{-(N+1)}$ and $10^{-(N-1)}$ as well, so you don't miss the best value.

4. Differentiate $f(x) = x^2 - x^{0.5}$ analytically, and then check the result numerically at the point $x = 7$. Be sure to show all steps of your work.

5. Use the chain rule to find dz/dx when $y = \sin bx$ and $z = y^2$.

6. As manager of a large water quality research project, you have to design some special sample bottles, of which several thousand will be required. Each must:

• be cylindrical in shape

• have 1 liter capacity (1000 cm^3)

• have the smallest internal surface area consistent with the other two criteria. This results from the need to line the bottles with a very costly non-reactive substance.

What should be the inside dimensions of the bottles, to satisfy these criteria?

7. You are setting up an experiment in a greenhouse. You need a seed bed surrounded by a plant-free border of 10 cm at the top and bottom, and 6 cm on each side. Space is limited, and you have been allocated a *total* of 2000 cm^2 of area (seed bed plus border) on one of the tables. What overall dimensions should you choose to obtain the maximum area of *seed bed*? What will the actual seed bed area then be? (Note: the border is not part of the seed-bed area.)

   As usual, try to solve this problem both symbolically and numerically. Be sure to define any symbols you use.

8. Suppose it is found that the average annual concentration [ppm] of a pollutant at a "target point" at least 1 km away from a pollutant source is proportional to the average annual pollutant concentration at the source, divided by the distance between the source and the target point. The proportionality coefficient is some constant k [km]. Now consider a small city with two major sources of SO_2. The average SO_2 concentration at one is 110 ppm, and at the other is 230 ppm. The two sources are 7 km apart. If their effects at a target point are additive, and all other sources can be neglected, where along the line between the two sources (but at least 1 km from each) would the pollutant be a minimum?

9. Suppose the white oaks in a forest are a fraction f of the individual trees there, and that there are a total of n_t trees ha^{-1} (all species) in that stand. In other words, the stand contains $f \cdot n_t$ white oaks per hectare. Under that condition, the oaks produce m kg of acorns per tree per year, on average. Suppose that if the density of trees (stems per hectare) increased, the acorn production of each tree would decrease by δ kg per tree per year per added tree [kg tree^{-2} yr^{-1} overall].

   A. If only white oaks were added (and no stems of other species) what number of white oaks per hectare would produce the largest number of acorns per hectare per year? (The deer—weeds of the mammal world—who eat acorns, would like to know this.) For this part, provide that answer entirely in symbols.

   B. If $f = 40\%$, $n_t = 120$, $m = 200$, and $\delta = 4$, what number of white oaks would lead to maximum production of acorns, and what would that production be?

Hint: For a given density x of white oaks per hectare, acorn production p (you determine the units) is reduced from its original value, m, by an amount $\delta(x - fn_t)$.

10. The Ostrich Waste Disposal Corporation (OWDC) plans to build a landfill. They are constrained by state regulations to lay it out in trenches 10m wide, but they have some choice about trench depth. The trenches have vertical sides, and a horizontal bottom. They need a total trench volume of 30,000 m³.

    The top covering for the trenches costs C dollars per square meter, and to reduce that cost, they would like to make the trenches deep. On the other hand, excavation costs increase quadratically with depth, and can be estimated as $E = a + bZ^2$, where E is the excavation cost per unit length of 10-m wide trench [dollars/m], Z is trench depth [m], and a and b are constants.

    A. What are the units of a and b?

    B. What length and depth should the company choose to minimize its total construction costs for a landfill of the necessary size? Work this part entirely in symbols. Express your final answer in one (or a few) complete sentences.

11. The equation

$$L = a + b \sin\left[\frac{2\pi}{c}(t + d)\right]$$

    is an approximate description of the daylength L between sunrise and sunset at 32 N latitude, as it varies with time of year t. (For data, see List 1949.) We'll work with 365-day years, and ignore leap years. The variables are:

    L = daylength [minutes]
    a = annual average daylength [minutes]
    b = amplitude [minutes]
    c = period [days]
    d = "phase shift" [days], to make the peak fall near June 21.

    We will define $t = 0$ at midnight between Dec. 31 and Jan. 1.

    Sketch a graph of L versus t and work out the following:

    • Differentiate the equation analytically (symbolically) to yield dL/dt.

- Using that derivative, calculate the numerical value of dL/dt at *noon* of January 1 (when $t = 0.5$ da) and at noon on September 21 (when $t = 263.5$ da). State the units and explain the physical meaning of these numbers. For these calculations, use $a = 731.5$, $b = 170.5$, $c = 365$, and $d = -80.5$.

- Check your derivative value for noon of January 1 by differentiating the original equation for L numerically at $t = 0.5$.

12. A forest ecologist estimates that the density D of acorns dropped near white oak trees decreases with distance x from the tree, with the relationship being $D(x) = a/(1+x)$, where D is in acorns m^{-2} and x is in m. However, deer and squirrels are aware of this distribution—they learn that relationship in school—and thus forage most intensely near the trees. As a result, the probability P of any given acorn remaining on the ground and germinating increases with distance from the tree as given by $P(x) = bx/(c+x)$. (That probability is also the fraction that germinates, on the average.)

    Your tasks are to:

    a) Determine the distance x_{max} where the density of *germinating* acorns (in number m^{-2}) is greatest. Your answer should be given in terms of the symbolic constants, a, b, and c.

    b) Explain how you know that you have found a maximum rather than a minimum.

    c) If $a = 3$ acorns m^{-1}, $b = 0.5$ m^{-1}, and $c = 40$ m, then what is the numerical value of x_{max} and of the maximum density of germinated acorns?

    You'll likely find it helpful to sketch the two functions, and perhaps some combination of them as well.

    Hint: Remember that a fraction can be zero either if the numerator is zero or if the denominator goes to infinity. In max-min problems, we are usually interested in solutions where the numerator goes to zero.

13. One afternoon while searching for spotted-owl nests, you use a topographic map and a GPS unit to keep careful track of where you are. At quitting time, you find yourself one mile east of the

north-south road on which you left your truck. Specifically, if you walked due west you would strike the road at a point three miles south of your truck.

You could walk the four miles along that right-angle path and you could walk straight toward your truck, as two limiting cases. However, you think you would get to your truck fastest if you angled through the woods to strike the road at a point that was less than three miles from the truck. You estimate that your walking speed through the woods would be two miles per hour, and your speed on the road would be four miles per hour.

a) To minimize your walking time, what point on the road (at what distance from your truck) should you aim for? For both parts of this problem, work in symbols for as long as you can, but then give numerical values.

b) What would your minimum walking time be, and how much time would you save compared with that along the right-angle path?

As always, you'll likely a sketch helpful.

Exercises on Taylor & Maclaurin Series

Reminder—In the field of analysis, one learns that "nice" functions (even transcendental ones[8]) can be expanded as Taylor series that are valid for all x within certain intervals. If a function $f(x)$ is "expanded about" a point a, it takes the form described by Eqn. 2.10. Such a form can always be simplified to an infinite polynomial of the form

$$f(x) = b_0 + b_1 x + b_2 x^2 + b_3 x^3 + \ldots$$

and this fact is useful for doing approximate calculations in numerical analysis.

14. Show that:

- $\cos(-a) = \cos(a)$. (Because of this, $\cos x$ is called an "even" function.)

---

[8] A transcendental function is one that can't be expressed as a ratio of two polynomials. Examples are e^x and $\sin x$.

- $\sin(-a) = -\sin(+a)$. ($\sin x$ is an odd function.)
- $e^{iz} = \cos z + i\sin z$, where $i = \sqrt{-1}$. To confirm this, it helps to work out first the values of i^2, i^3, i^4, etc. You will find that an interesting pattern emerges.

15. The purpose of this exercise is to show that the Taylor series for a finite polynomial *is* that polynomial. E.g., consider

$$f(x) = P_3(x) = 7x^3 - 3x, \text{ expanded about } a = 4. \qquad (2.16)$$

We have: $f(a) = 7(64) - 3(4) = 436$, and from the rules for derivatives of positive powers (i.e., that $dx^n/dx = nx^{n-1}$):

$$f'(x) = 21x^2 - 3, \text{ so } f'(a) = 21(16) - 3 = 333,$$

$$f''(x) = 42x, \text{ so } f''(a) = 42(4) = 168, \text{ and}$$

$$f'''(x) = 42, \text{ so } f'''(a) = 42.$$

Also, $f^{(n)}(x) = 0$ for all $n > 3$ and for all x. Substituting these into (1), we get the Taylor series:

$$f(x) = 436 + 333(x - 4) + \frac{168}{2 \cdot 1}(x - 4)^2 +$$

$$\frac{42}{3 \cdot 2 \cdot 1}(x - 4)^3 + 0 + 0 + \ldots.$$

Your jobs are to simplify this expression by grouping like powers of x, and to show that it reduces to the original polynomial (Eqn. 2.16).

16. Recalling that

$$\frac{d}{dx}(x^{-n}) = -nx^{-(n+1)},$$

expand $y = 1/x$ about the point a=2. Show the first 4 terms of the expansion. Use this truncated series to estimate $f(2.1)$, and compare the estimate with the true value of $f(2.1) = 1/(2.1)$ obtained by division.

Note that the function $f(x) = 1/x$ is *not* a polynomial because it is not a sum of terms of *positive* powers of x. Hence, to be *exact* the Taylor series for this function must have an infinite number of terms. Do you think your series would yield a reasonable approximation for $f(x)$ at $x = 0$? Why or why not?

17. Expand $y = R + Sx + Tx^2$ about $x = a$ (symbolically), and show that the Taylor series expansion reduces to the original polynomial.

18. Given: The derivative of $y = e^{bx}$ is be^{bx}, and (as you may remember) for *any* function y and *any* constant c,

$$\frac{d(cy)}{dx} = c\frac{dy}{dx}.$$

Using these facts, expand $y = ce^{bx}$ as a Maclaurin series (which means you set $a = 0$). Then show that that series reduces to c times e^u when $u = bx$. For this, you can take as given that for any u,

$$e^u = 1 + u + \frac{u^2}{2!} + \frac{u^3}{3!} + \frac{u^4}{4!} + \ldots.$$

19. Expand $f(x) = xe^x$ about $a = 1$ through the $[x - 1]^3$ (third-order) term. What are the relative errors that result if you use this truncated Taylor series to estimate $f(1.1)$, $f(1.2)$, $f(1.4)$, $f(1.8)$, $f(2.6)$, $f(4.2)$ and $f(7.4)$?

20. Expand the function $f(x) = \log(x)$ as a Taylor series around the point $a = 1$, keeping terms up to and including the term based on the third derivative.

What approximate value does your series provide for $\log(1.1)$? Then, if you assume that your calculator yields the exact true value for that quantity, what is the relative error of the approximate value calculated from your series?

21. The net exchange R of long-wave (thermal) radiation between an object (such as a leaf or an animal) and its surroundings is given by $R(T) = \sigma\epsilon(T^4 - T_s^4)$, where σ is the Stefan-Boltzmann constant [J cm^{-2} s^{-1} deg^{-4}], ϵ is the emissivity (or "blackness") of the object in far-infrared wavelengths [unitless], T is the object's absolute

surface temperature [deg K], and T_s is the absolute temperature of surrounding objects. R is one term of the energy balance of an object. Assuming that T_s is a known constant, determine the first five terms of the Taylor series for $R(T)$, expanded about the temperature $T = a$.

22. This problem makes use of the Maclaurin series for e^x, which as you have seen is

$$e^x = 1 + x + \frac{x^2}{2!} + \frac{x^3}{3!} + \cdots . \tag{2.17}$$

In risk assessment for carcinogens, test animals like rats are often fed high doses of a chemical. Then assessors use a mathematical model to estimate the mean number μ of tumors per animal expected to occur in rats fed some lower dose to which humans might be exposed. From μ, the assessors wish to estimate the *carcinogenic potency* ϕ, which is defined as the probability that a rat fed that lower dose would get cancer; i.e., one or more tumors.

At the low doses considered, cancers are rare, and so the Poisson distribution from statistics can be used to show that the probability of a given rat's *not* getting cancer is $P(0) = e^{-\mu}$. Thus the probability that a rat will suffer from one or more tumors is $\phi = 1 - P(0) = 1 - e^{-\mu}$, and this is the quantity we seek. Suppose for a particular dose of a particular chemical that μ is quite small, say 0.0001. Then use the series in Eqn. 2.17 to show that for all practical purposes, $\phi = \mu$. Describe your logic. (Ultimately the potency for rats is converted to a potency for humans, but that's another story.)

23. (This problem is *about* integration, but does not require you to perform any integration.) As noted on p. 44, the equation for the "bell curve" of statistics; i.e., for the standard normal distribution, is

$$p(z) = \frac{1}{\sqrt{2\pi}} \exp\left(-\frac{z^2}{2}\right). \tag{2.18}$$

Unfortunately that analytic expression is not directly useful much of the time because it yields the probability *density* at any value of z, and not the actual probabilities that we usually want to work with. To calculate the probability P that a standard normal random variable Z lies in some range of interest, we have to integrate $p(z)$ over that range. Sadly, $p(z)$ can't be integrated analytically.

Suppose we needed $P(0 \le z \le 0.8)$ for some statistical application. One way to obtain that value would be to write the Taylor series for $p(z)$ (expanded about $a = 0.4$, say), to truncate that series after k terms, and to sum the integrals of those terms. If we kept enough terms, that should work reasonably well, since the integral of a sum is the sum of the integrals.

Although you need not perform any integration, your task here is to check out how well a few terms of the Taylor series for $p(z)$ approximate that function. In particular,

A. Find the first three terms (i.e., through the quadratic term) of the Taylor series for $p(z)$, expanded about 0.4.

B. Calculate the true numerical value of $p(0)$ and of $p(0.8)$ from Eqn. 2.18, to as many digits as your calculator supplies.

C. Calculate the approximate value of $p(0)$ that would be obtained from the three terms of the Taylor series, and determine the relative error of that approximation.

D. Repeat the calculations from Part C at $z = 0.8$.

If those errors are relatively small, that would suggest (but not prove) that the integral of the three-term series would approximate the integral of the true $p(z)$ reasonably well over this range. If not, you might want to add more terms, but as the higher derivatives get messy, we'll omit that here.

Hints: To simplify developing the Taylor series, you may find it helpful to give the leading constant in the formula for $p(z)$ a symbolic name (like c); use a substitution like $g(z) = -z^2/2$, and make use of the chain rule.

24. At $x = 3$, a certain function has the value $f(3) = 1$, and its first three derivatives there have the values $f' = 1/2$, $f'' = 1/4$, and $f^{(3)} = 1/8$. The fourth and all higher derivatives are zero at $x = 3$. Using this information and the properties of Taylor series, write a formula that would allow an assistant to calculate values of $f(x)$ for any value of x.

25. In energy-balance calculations similar to the one for the top of a car on p. 8, a quantity like $q(T) = (T^4 - A^4)$ frequently arises, and analysts often "linearize" that with the help of Taylor series to simplify solving for temperature. Find the first three terms of the Taylor series for the function $q(T)$ expanded around the point $T = A$; i.e., the constant term and those involving $(T - A)$ and $(T - A)^2$. Then, calculate the numerical values of each of those three terms if $T = 300°$ K and $A = 295°$ K. Linearization involves dropping the third (and higher) terms. How large is that third term, relative to the sum of the first two, in this application? Can it reasonably be dropped?

2.7 Questions and Answers

1. Please explain again why the derivative of $y = f(x) = e^{bx}$ is $f'(x) = be^{bx}$.

   • This is a good place to use the chain rule. Define $u = bx$. Then $y = g(u) = e^u$, and $u = h(x) = bx$. The chain rule says that $dy/dx = (dy/du)(du/dx) = (e^u)(b)$. Now replace the u with bx, and you have $dy/dx = (e^{bx})(b) = be^{bx}$. Alternatively, you can work with the Maclaurin series for e^u; i.e., with

$$e^u = 1 + u + \frac{u^2}{2!} + \frac{u^3}{3!} + \frac{u^4}{4!} + \dots .$$

   Now substitute $u = bx$, and then differentiate the result term by term. Remember that the derivative of a sum is the sum of the derivatives. You should be able to factor out a b from the result, and what's left will be $\exp(bx)/b = \exp(u)$.

2. Is there a table of derivatives and integrals we can use for reference?

   • There are lots available. The firm that publishes the *Handbook of Chemistry and Physics* extracts the math tables from that book as a smaller book, and you can buy that. The *Handbook of Mathematical Functions* is a fine reference book by Abramowitz and Stegun, originally published by the National Bureau of Standards (Now the National Institute of Standards & Technology), and later, in paperback, by Dover—it's a big tome, however. Some of the "outline series" of books (e.g., Schaum's) have tables available, I think. I've seen other books of tables in various bookstores, and any of them ought to be helpful.

3. I'm unsure of when you use $[f(a + h) - f(a)]/h$ and when you just differentiate analytically.

• You should differentiate analytically most of the time—it's more exact. But you can use the finite difference form (A) to *check* analytic derivatives, (B) when you have only a list of values of the function, not the function itself, and (C)—yet to come—as the basis for approximate solutions to differential equations. Remember that the central difference form $[f(a + h) - f(a - h)]/(2h)$ is almost always preferable to the forward difference you asked about.

4. Please give another example of the chain rule.

• That would also be a good thing to look up in a calculus text for a good general description. For an example, though, suppose the water temperature in some lake varied with time on summer days as $T = a + b\sin(ct)$, and suppose evaporation rate from the water depended on temperature according to $E = \alpha \cdot \exp(kT)$. Then ultimately E would vary with time through this "chain" of dependencies. Thus, to see how rapidly E would change with time at some particular time, you could use $dE/dt = (dE/dT)(dT/dt)$, which is the chain rule. Here $dE/dT = \alpha k \exp(kT)$, and $dT/dt = bc\cos(ct)$.

5. Why again is the derivative of $\log(bx)$ equal to $1/x$?

• $\log bx = \log b + \log x$ (log of a product is the *sum* of the logs). $\log b$ is a constant, so its derivative is zero. Thus, $d(\log bx)/dx = 0 + d(\log x)/dx = 1/x$. That one seems counterintuitive at first, doesn't it?

6. When you gave the rule that $d(u+w)/dx = du/dx + dw/dx$, why did you specify that $u = f(x)$ and $w = g(x)$?

• It only makes sense to differentiate u with respect to x if u is a function of x. The '$u = f(x)$' was just meant to tell you that it is.

7. What are the applications of Taylor series?

• They are often used as the basis for approximations.

They form the basis for much of numerical analysis. (This is related to the first point.) We will use them at various times in the semester either to derive numerical methods (e.g., Newton's method for finding roots of non-linear equations) or to justify and help understand the behavior of other methods.

They are used a great deal by mathematical modellers, engineers, and other applied mathematicians. For those of you whose goal in the course is to be able to work with such people, a basic understanding of Taylor series (and that's our level—very basic) should help you communicate with them.

A side benefit in working with Taylor series is that you get some useful practice differentiating.

8. How can you get so much information from a Taylor series, when all you know about is one point?

• Well, you have to know a lot about that one point! You'll see as we get further into it (the short distance farther that we *will* go) that the first N terms of a Taylor series are equivalent to a polynomial of order $N - 1$. That means that knowing the function and N derivatives at a single point tells you quite a lot about how the function curves and varies, at least in the near neighborhood of that point.

If you know the y value at a given x, and you know the slope, you can move left or right a short distance on the tangent line, and be close to the value of the function there. If you know both the first and second derivatives there, that gives you an estimate of how much you should correct for curvature. Each additional derivative helps you to correct more and more.

9. In a Taylor series, what are a and x, and especially, what is the difference?

• x is the general variable (on the x axis), while a is the specific value of x about which we "expand the function." That is, a is the x value for which we must know the value of the function and its derivatives. If we truncate the series at a finite number of terms, then the resulting approximation will generally be better when x is near a than when x is far from a.

10. If we can obtain square roots from our calculators, or if we already know some other $f(x)$, why do we need Taylor series?

   • The square root example illustrated what a Taylor series is, the steps you have to go through to get one, and the fact that these series yield better approximations near a (i.e., in a narrow range around the point of expansion), and better approximations with more terms (at least near a).

   If you go on a long research trip some time, and drop your calculator in a lake, you now know one way to get the square root of 5 <grin>. That's not *too* far fetched, I guess. You might have to triangulate with $x^2 + y^2 = z^2$ $\left(\text{from which } z = \sqrt{x^2 + y^2}\right)$, to find locations on a lake, or some such thing.

   More realistically, If your calculator doesn't give hyperbolic sines and cosines, you now have a way to get those. That is, Taylor series are used for many other kinds of functions in addition to square roots.

   Most important, we now have a justification for using polynomials to approximate general (non-polynomial) functions, and we know that these approximations work better in narrow ranges and with more terms. This is the major reason we deal here with Taylor series—they form the basis for most numerical methods.

11. Does the second derivative show how the slope is changing? For example, does a negative second derivative tell you that the slope is getting less steep as x increases?

   • Yes, and yes. The second derivative is just the slope of the slope (It tells you how fast the *slope* changes with increasing x). A negative f'' tells you that f' gets smaller as x increases.

   When you do a max-min problem, you find a point where the first derivative is zero. To determine whether that is a maximum or a minimum point, you can calculate the second derivative at that point. If f'' is positive there, it means that the slope keeps getting bigger as you move to the right. Therefore, the function is cupped upward, and you have a minimum. If f'' is negative at your extreme point, then the slope keeps getting smaller as you move to the right. That means that the function is cupped downward at the extremum, so the point is a maximum. See any calculus book if you want more on this. Also, try some sketches to see it better.

Conceptually, this indicates why including a second-derivative term $(f''(x - a)^2/2!)$ improves a Taylor series. The information about how the slope changes as you move away from $x = a$ helps to predict $f(x)$, relative to what you would get from using just $f(a)$ and $f'(a)$ (the starting value and the starting slope).

12. How is a Taylor series different from linear approximation, where $f(x) \approx f(a) + f'(a)\Delta x$? Is the Taylor series more exact?

 • One way to look at this is that a linear approximation is the simplest form of approximation based on the Taylor series—it is equivalent to using the first two terms of a Taylor series. The Taylor series is more exact if you keep more terms, but if you drop all but the first two terms, the two are identical. Linear approximations are very common, but during the semester we will use some quadratic ones (like the central difference derivative) and one quartic (fourth-order) one. Stay tuned.

13. What does your example on p. 35 have to do with Taylor series? What do those calculations illustrate?

 • Those calculations are in the "numerical differentiation" section. They are indirectly related to Taylor series in the sense that Taylor series tell us it is better to use central differences than to use forward differences. But these calculations illustrate a completely different point, namely that if you use too small an "h" value (that would be good from the point of view of Taylor series, which tell us to work in a very small range), then the new problem of round-off error rears its ugly head. There are tradeoffs like this in a lot of numerical work. Keeping ranges small helps from one point of view, but often increases round-off error. Applied math (like life) is full of compromises.

14. Shouldn't there be a 'remainder term' at the end of a Taylor series?

 • If you truncate a Taylor series (i.e., cut it off after a finite number of terms) then it would often have a remainder term added to it, as you have evidently seen elsewhere. However, so far I think the series I've written have all had a "$+\ldots$" at the end, indicating a (theoretically) infinite series of terms. The full series does not have a remainder. Also, Taylor series for polynomials can be exact with a finite number of terms, and they don't need remainder terms either.

15. What is the meaning of second and higher derivatives, like $f''(x)$, $f'''(x)$, and so on?

    • That would be a good thing to review in an introductory calculus text, but let me explain it too. Take the function $y = f(x) = x^5 + 3x$, say. Wherever you go on the x axis, the slope of the curve above that point (the "rise" over the "run") can be calculated using the first derivative, which is $f'(x) = 5x^4 + 3$. You can treat that derivative as a function too. If you do, and take *its* derivative everywhere, you would get "derivative of $5x^4 + 3$" $= 20x^3 + 0 = 20x^3$. That new function is just the *second* derivative of the original function. It tells you how fast the slope of the original function changes as you move from one value of x to another. A 20th derivative is just the derivative with respect to x of the 19th derivative (etc.). (For $f(x) = x^5 + 3x$, every derivative above the fifth would be zero, however.)

16. What is the point of working with a Taylor series if you just get the original function back?

    • That happens only if your original function is a polynomial. It didn't happen with the square-root example we've started to work with, as you saw. Finding the Taylor series for a polynomial is instructional, though, because it shows the close relationship between those two mathematical forms.

17. With a Taylor series, do you always need an infinite number of terms, or can you stop once the terms become zero?

    • (Infinity) × (zero) is still zero, so *if* you know that all remaining possible terms are zero, then you can stop because you know that adding them won't make any difference. Generally that happens only with polynomials, however.

Chapter 3

Integration

3.1 What is Integration?

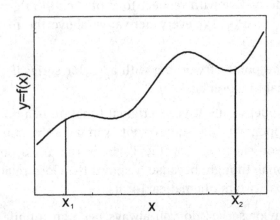

Figure 3.1: A function for defining the definite integral.

If $y = f(x)$ is the function plotted in Fig. 3.1, what is the meaning of the definite integral given by

$$I \stackrel{\text{def}}{=} \int_{x_1}^{x_2} y \, dx = \int_{x_1}^{x_2} f(x) \, dx?$$

A geometric interpretation would be "the area under the curve over the range $x_1 \le x \le x_2$." A more mathematical answer involves breaking the area into numerous vertical panels of width h_i and height $f(x_i)$, where x_i is a value anywhere in the range of the panel. Then the integral is defined as[1]

$$I = \lim_{\substack{n \to \infty \\ h_i \to 0 \, \forall i}} \left[\sum_{i=1}^{n} h_i f(x_i) \right].$$

Adding an infinite number of infinitesimally narrow panels in this way yields the total area.

Important note: expressions like $\int \sin x$ or $\int (a + bx + ce^{3x})$ are meaningless, and have incorrect units as well. Every "$\int$" sign must be accompanied by "d something" to have any meaning at all. Please be sure that every integral sign you write is accompanied by a differential of some variable.

3.2 Usefulness in Environmental Science

Integration is mathematically equivalent to adding an infinite number of infinitesimal parts; as a result, it usually arises in situations involving accumulation and averaging.

Example 1. A major application is to solve differential equations. E.g., let R_D be the rate of dumping (in kg mo^{-1}) of PCB mass M at a landfill. Suppose R_D rose linearly with time starting in June 1962 (which we define as $t=0$). In particular,

$$R_D = 500 + 50t \text{ kg/mo} = \frac{dM}{dt} = a + bt.$$

To determine how much total PCB mass had been dumped t months later, we integrate that variable rate of dumping over time:

$$M = \int_0^t R_D \, dt = \int_0^t (a + bt) dt = a \int_0^t dt + b \int_0^t t \, dt =$$

$$a(t - 0) + \frac{b(t^2 - 0^2)}{2} = 500t + \frac{50}{2}t^2.$$

A unit check yields kg = [kg mo^{-1}][mo]. This is a good example of integration applied to a process of accumulation, in this case accumulation of PCB mass.

Example 2. Consider a big cat that is 50 m away from an antelope (Fig. 3.2, p. 54). Each animal has a maximum speed V (V_A or V_c)

---

[1]The symbol "$\forall$" means "for every."

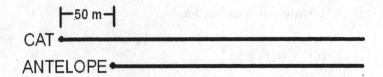

Figure 3.2: Starting positions and paths of the cat and the antelope.

in m/s. A reasonable model for changes in velocity resulting from acceleration might be $v(t) = V(1 - e^{-t/\tau})$ for each animal; τ here is called a "time constant" (about which more later).

Suppose the antelope has a greater maximum speed ($V_A > V_C$), but the cat can accelerate faster; hence has a smaller τ. The numerical values might be:

	V	τ
Cat	8.5 m/s	2.5 s
Antelope	9 m/s	5 s

For these values, will the antelope be caught? To answer this question mathematically, we need to know how far each animal will go in t seconds, and determine whether the cat would ever have gone 50 m farther at any time. Because $v = dx/dt$—velocity is the rate of change of distance with time—that distance is the integral of the velocity; i.e., for each animal[2]

$$x(t) = \int_0^t v(t)\, dt = \int_0^t V(1 - e^{-t/\tau})\, dt = V(t + \tau e^{-t/\tau} - \tau) \quad (3.1)$$

Fig. 3.3, p. 55, shows the distance between the cat and the antelope, calculated as $50 - x_{cat} - x_{antelope}$, as that distance varies through time. Because that distance never goes to zero, the antelope must escape.

Again, integration often models accumulation, and here it models accumulation of distance through time.

Example 3. A tank initially contains a volume V_0 m^3 of water. Water flows in at a (possibly variable) rate, $q_{in}(t)$ [m^3 hr^{-1}], and out at a rate $q_{out}(t)$ [m^3 hr^{-1}]. The volume V in the tank varies with time according to $V(t) = V_0 + \int_0^t (q_{in} - q_{out})\, dt$. This process represents a good physical model for the meaning of integration.

---

[2]We haven't reviewed the mechanics of integration yet, so don't worry if you don't see how this result arose. We'll take up those operations soon.

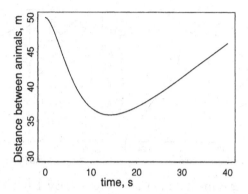

Figure 3.3: Distance between the cat and the antelope through time.

Integration and Averaging

Consider the dissolved oxygen curve shown in Fig. 3.4. If the only data one had were "grab sample" values taken every two hours during the day (at 0, 2, 4, ..., hours), then the average DO concentration for that day could be estimated as

$$\bar{c} = \frac{1}{13} \sum_{i_1}^{13} c_i,$$

an ordinary-looking average. Actually, a better value in this case would be $(c_0 + 2c_2 + 2c_4 + \ldots + 2c_{22} + c_{24})/24$. That *weighted mean*

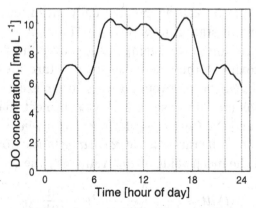

Figure 3.4: Concentration of dissolved oxygen (DO) in a stream over a day. The curve represents the results of continuous monitoring; alternatively, grab samples might be taken every two hours.

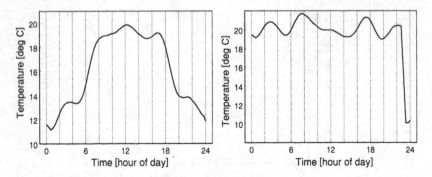

Figure 3.5: Temperature variation at a site on Day 1 (left), and on Day 2 (right).

would account for the first and last values each representing only one hour of the day under consideration, while the other eleven values each represent two hours.

If one had the continuous record shown in the figure, and defined that curve as $c(t)$, then the *true* mean could be calculated as

$$\bar{c} = \frac{1}{24 - 0} \int_0^{24} c(t)\, dt.$$

This illustrates the general definition that the mean of a continuous function $f(x)$ for x between two limits a and b is given by

$$\text{mean} \stackrel{\text{def}}{=} \int_a^b f(x)\, dx / (b - a).$$

The US weather bureau defines the daily mean temperature at a given measurement station to be $(T_{max} + T_{min})/2$; i.e., halfway between the highest and lowest temperatures over the day. How well does this approximate the true mean temperature for the two days shown in Fig. 3.5? How well would it work for an exact sine wave?

A ratio problem[3]: Suppose in a 10-year project that the Benefit = \$10 M/yr=a and the Cost = (\$10M)($e^{-0.23t}$)/yr =be^{-ct} as shown in Fig. 3.6, p. 57. Then the ratio *at any particular time t* would be $R(t) = B(t)/C(t)$, and the mean value of that ratio over the ten years is:

$$\bar{R} = \frac{1}{10} \int_0^{10} R(t)\, dt = \frac{1}{10} \int_0^{10} \frac{a}{be^{-ct}}\, dt = \frac{a}{10bc}(e^{10c} - 1) = 3.902.$$

[3]Be cautious with ratios—they can cause lots of trouble if you don't think carefully about how you use and interpret them.

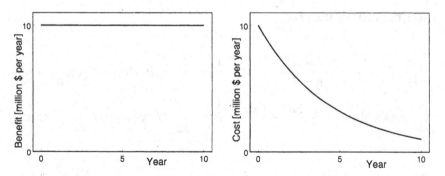

Figure 3.6: Benefit attained (left) and cost incurred (right) for an environmental protection project over a ten-year period.

However, $\bar{R}$, the average value of the ratio over time, isn't usually of much interest. What we really care about is the overall benefit:cost ratio defined by Eqn. 3.2.

$$\bar{R}_{\text{overall}} = \frac{\text{total } B}{\text{total } C} = \frac{\int_0^{10} B(t)\, dt}{\int_0^{10} C(t)\, dt}. \tag{3.2}$$

That has the value

$$\bar{R}_{\text{overall}} = \frac{10a}{\int_0^{10} be^{-ct}\, dt} = \frac{10ac}{-b(e^{-10c} - 1)} = 2.56,$$

which is substantially different from the mean of the ratio R over the period. Do think carefully whenever you use and interpret ratios!

3.3 Analytic Integration

Many, but by no means all, integrals can be evaluated analytically. That is, the result can be written symbolically in terms of standard functions, with a finite number of terms. Here are some integrals you should memorize, so as not to waste time looking them up every time you need one. Note that these are written as definite integrals, so we end up with a value at the upper limit minus a value at the lower limit. (With *indefinite* integrals, we have a value plus a constant instead.)

Integrals to Memorize

$$\int_a^b x^n \, dx = \frac{b^{n+1} - a^{n+1}}{n+1}; \quad \int_a^b \frac{1}{x} \, dx = \log b - \log a = \log \frac{b}{a}; \quad (3.3)$$

$$\int_a^b \exp kx \, dx = \frac{\exp(kb) - \exp(ka)}{k}; \quad \int_a^b cf(x) \, dx = c \int_a^b f(x) \, dx;$$

$$\int_a^b \sin kx \, dx = \frac{\cos ka - \cos kb}{k}; \quad \int_a^b \cos kx \, dx = \frac{\sin kb - \sin ka}{k};$$

$$(3.4)$$

$$\int_a^b [f(x) + g(x)] \, dx = \int_a^b f(x) \, dx + \int_a^b g(x) \, dx;$$

$$\int_a^b [f(x) \cdot g(x)] \, dx = ? \quad (3.5)$$

The integral in Eqn. 3.5 must be left unspecified; there is no general "product rule" for integrals comparable to the one for derivatives. However, integrals of some products will yield to *integration by parts*, for which see any calculus text.

I have emphasized definite integrals above because I find them most useful in applications. You will sometimes also see indefinite integrals, which for example take the form

$$\int x^n \, dx = \frac{x^{n+1}}{n+1} + C,$$

where C is a constant of integration that must be determined from the details of an application.

To perform analytic (symbolic) integration in MATLAB®, when its Symbolic Math Toolbox is available, you could enter (again, just the parts to the left of the dots):

```
syms a b c x.......................Define symbolic variables
f=sin(c*x) .........................Define the function to integrate
int(f) ...............................Find indefinite integral
int(f,a,b)..........................Find definite integral
```

For the indefinite integral, MATLAB returns ans=-1/c*cos(c*x), and for the definite one, ans=-(cos(c*b)-cos(c*a))/c. Note well that MATLAB does *not* provide a constant of integration for symbolic *indefinite* integrals. You must remember to add your own.

As with differentiation, I suggest that you try using MATLAB to obtain the other integrals in Eqns. 3.3–3.4 above.

Integration by Substitution

Many integrals not in that group will yield to substitution. For example, to integrate

$$\int_{t_1}^{t_2} \exp[a + b(t-1)]dt,$$

let $u \triangleq a + b(t-1)$. Then the integral becomes

$$\int_{t_1}^{t_2} \exp(u)dt = \int_{t_1}^{t_2} e^u dt.$$

However, $du/dt = b$, so $du = b\,dt$, and our integral becomes[4]

$$\frac{1}{b}\int_{t=t_1}^{t=t_2} e^u du = \frac{1}{b}\Big[\exp(u)\Big]_{t=t_1}^{t=t_2}$$

$$= \frac{1}{b}\Big\{\exp[a + b(t-1)]\Big\}_{t=t_1}^{t=t_2} \qquad (3.6)$$

$$= \frac{\exp[a + b(t_2-1)] - \exp[a + b(t_1-1)]}{b}$$

There are other "tricks" for integrating functions, such as integration by parts. Because such most integrals can be more easily obtained using tables or software, we will not take the time to review those methods here.

Integration Using Tables

One of the easiest analytic integration methods is to "look it up in the tables!" My favorites are the *CRC Standard Mathematical Tables*

---

[4]When you change the variable of integration in association with a substitution, it is helpful to keep explicit track of the limits of integration, as with the "$t = t_1$" and "$t = t_2$" in Eqn. 3.6.

Table 3.1: A short table of integrals. You are likely to find others in calculus texts, and much more complete tables are available in various handbooks.

$$\int e^{ax} dx = \frac{e^{ax}}{a} + C$$

$$\int \sin ax \, dx = \frac{-\cos ax}{a} + C$$

$$\int \cos ax \, dx = \frac{\sin ax}{a} + C$$

$$\int \sin^2 ax \, dx = \frac{-1/2 \cos(ax) \sin(ax) + 1/2 \, ax}{a} + C$$

$$\int \cos^2 ax \, dx = \frac{1/2 \cos(ax) \sin(ax) + 1/2 \, ax}{a} + C$$

$$\int \log ax \, dx = x \log(ax) - x + C$$

$$\int \frac{\exp(ax)}{1 + \exp(ax)} dx = \frac{\log(1 + e^{ax})}{a} + C$$

(extracted from Lide 2005), and the *Handbook of Mathematical Functions* (Abramowitz and Stegun 1972). These days, however, it may be even faster to use software that can integrate analytically. A limited table is provided as Table 3.1, and some others that involve factors like e^{ax} are provided in a second table on p. 131.

3.4 Numerical Integration

Although many integrals can be evaluated analytically (i.e., symbolically, in closed form), many other important ones cannot be. Yet, as users of math, we may still need to know the value of such integrals. For example, users of statistics frequently need to know integrals like

$$P(z) = \frac{1}{\sqrt{2\pi}} \int_{-\infty}^{z} e^{-z^2/2} dz = \int_{-\infty}^{z} p(z) \, dz, \qquad (3.7)$$

since these give the cumulative probability lying below z in the unit normal distribution. Because of their importance, some values of this integral are generally available in tables. The tabled values, however, come from numerical methods provided by the field of *numerical analysis.* Because we need numerical methods for many integrals, we take up that topic next.

Suppose, then, that we wished to find

$$I = \int_a^b f(x)\, dx.$$

In a general sense, all numerical integration methods involve some scheme to estimate the area under the curve defined by the function of interest. Our brief study of Taylor series suggested approximating functions by polynomials, and many schemes for numerical integration do just that — a polynomial is fit to some set of points along the curve [5], and the integral of that polynomial provides an approximation to the area under the curve.

The simplest version of that scheme, the *trapezoidal rule*, fits a straight line between the points $[a, f(a)]$ and $[b, f(b)]$, and uses the area under the resulting trapezoid as the approximate integral. A more refined scheme, known as *Simpson's 1/3 rule*, fits a quadratic to those two endpoints plus the point halfway along the range of x—we consider that method in more detail shortly.

Other methods of "quadrature" use higher order polynomials, which in effect use more terms of the Taylor series for the function; however, care is required because high degree polynomials can take wild swings, and deviate far from the true curve between the points being fit. Other more complex approaches exist as well, but we will deal with the simple and effective Simpson's rule, which works as follows.

Suppose we wished to integrate a function, like the one plotted as a solid line in Fig. 3.7, p. 62, from $x = x_1$ to $x = x_2$. In principle, $f(x)$ could be expanded in a Taylor series, and over a narrow enough range x_1 to x_2, we could approximate the Taylor series with a quadratic, as shown by the dashed line in the figure. That is, we would take

$$f(x) = a + bx + cx^2 + dx^3 + \ldots \approx a + bx + cx^2.$$

The middle expression there is the full Taylor series, and the RHS (right-hand side) is a quadratic approximation of it. In one application of the basic form of Simpson's rule, we find the approximate area of two *panels*, each of width h, as shown in the figure. Thus,

---

[5] Any n distinct points determine a polynomial of degree $n - 1$.

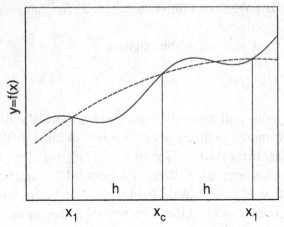

Figure 3.7: A function to be integrated approximately (solid line), with the quadratic approximation to be integrated by Simpson's rule (dashed line. Two *panels* of width h are shown.

$$I \approx I_a \stackrel{\text{def}}{=} \int_{x_1}^{x_2} (a + bx + cx^2)dx.$$

Now we define $x_c \stackrel{\text{def}}{=} (x_1 + x_2)/2$ and ask, can we get I_a into the form $I_a = k_1 y_1 + k_c y_c + k_2 y_2$, where y_c is the function value at the midpoint (x_c) between x_1 and x_2? That is, can we find constants k_1, k_c, and k_2 so that is true? Try it. First integrate I_a analytically:

$$I_a = \int_{x_1}^{x_2} (a + bx + cx^2)dx = \left[ax + \frac{1}{2}bx^2 + \frac{1}{3}cx^3 \right]_{x_1}^{x_2}$$

$$= a(x_2 - x_1) + \frac{1}{2}b(x_2^2 - x_1^2) + \frac{1}{3}c(x_2^3 - x_1^3).$$

That is the exact integral of the quadratic, but not necessarily of the real function. Note that the x-dependence in these three terms can be rewritten as:

$$(x_2 - x_1) = 2h,$$

$$(x_2^2 - x_1^2) = (x_2 - x_1)(x_2 + x_1) = 2h(x_2 + x_1), \text{ and}$$

$$(x_2^3 - x_1^3) = (x_2 - x_1)(x_2^2 + x_2 x_1 + x_1^2) = 2h(x_2^2 + x_2 x_1 + x_1^2).$$

Thus

$$I_a = 2h \left[a + \frac{1}{2}b(x_2 + x_1) + \frac{1}{3}c(x_2^2 + x_2 x_1 + x_1^2) \right].$$

Now divide the $2h$ by 6, and multiply the expression in brackets by 6. Then

$$I_a = \frac{2h}{6}\left\{6a + 3b(x_2 + x_1) + 2c(x_2^2 + x_2 x_1 + x_1^2)\right\} \stackrel{\text{def}}{=} \frac{h}{3}\{\ \}$$

Will this fit into the form $\{\ \} = k_1 y_1 + k_c y_c + k_2 y_2$? We use the relationship $y_1 = a + bx_1 + cx_1^2$ and similar ones for $y_c(x_c)$ and $y_2(x_2)$. Then we separate out the terms in $\{\ \}$ so that $\{\ \} =$

$$a + bx_1 + cx_1^2$$

$$+4a + 2b(x_1 + x_2) + c(x_1^2 + 2x_1 x_2 + x_2^2)$$

$$+a + bx_2 + cx_2^2.$$

The first and third expressions here are y_1 and y_2, respectively, so we have shown that

$$\{\ \} = y_1 + [4a + 2b(x_1 + x_2) + c(x_1^2 + 2x_1 x_2 + x_2^2)] + y_2.$$

We can now set $k_1 = k_2 = 1$. Then, if we can massage the middle three terms into the form $k_c y_c$, we'll be home free. Note that $x_c = (x_1 + x_2)/2$ and that $(x_1^2 + 2x_1 x_2 + x_2^2) = (x_1 + x_2)^2$. We now rewrite $[\ ]$ as

$$4a + 4b\left(\frac{x_1 + x_2}{2}\right) + c(x_1 + x_2)^2 = 4a + 4bx_c + 4c\left(\frac{x_1 + x_2}{2}\right)^2$$

$$= 4a + 4bx_c + 4cx_c^2 = 4y_c.$$

Thus $\{\ \} = y_1 + 4y_c + y_2$! Finally, then, our approximate integral is

$$I_a = \frac{h}{3}(y_1 + 4y_c + y_2) = 2h\frac{(y_1 + 4y_c + y_2)}{6}. \qquad (3.8)$$

As stated earlier, Eqn. 3.8 provides an exact value of the integral for any quadratic and for any h. Note that the fraction on the RHS here is a weighted mean of y values, in which the value at the center of the interval is counted more than the values at the ends of the interval. The $2h$ part corresponds to the range of the integration (the length along the horizontal axis), while $(y_1 + 4y_c + y_2)/6$ is a (weighted) average height of the values the function takes on over that range.

The weighting counts the value in the middle four times as much as the values at the two ends.

Simpson's rule can be generalized. For example,

$$\int_a^b f(x)\,dx \approx \left(\frac{b-a}{2\cdot 3}\right)\left[f(a) + 4f\left(\frac{a+b}{2}\right) + f(b)\right] \qquad (3.9)$$

is a two-panel numerical approximation to the integral of any function $f(x)$ from a to b. However, if the interval $b - a$ is too wide, and if $f(x)$ is not well approximated by a quadratic[6], then the approximation to the integral may not be very good either. Actually, Simpson's 1/3 rule with two panels gives an exact result over *any* interval $b - a$ if $f(x)$ is any *cubic* polynomial. Proving that fact makes an interesting exercise.

Suppose we want to integrate a function $f(x)$, not a cubic, over some fairly wide interval. We could go two ways—one would be to use higher-order polynomial approximations, and many *quadrature* methods do this. We could also apply Simpson's one-third rule over more panels, and that is the approach taken here.

Let's rewrite Eqn. 3.9 as

$$\int_a^b y\,dx = \int_a^c y\,dx + \int_c^b y\,dx,$$

which matches common sense if you interpret integrals as areas under curves. So if we set c at the midpoint between a and b, we can integrate $f(x)$ from a to c with two panels, and from c to b with another two panels. We have:

$$\int_a^c y\,dx \approx \frac{h}{3}\left[y_0 + 4y_1 + y_2 \qquad\qquad\right]$$

$$\int_c^b y\,dx \approx \frac{h}{3}\left[\qquad\qquad y_2 + 4y_3 + y_4\right].$$

When we add these two areas, we get

$$\int_a^b y\,dx \approx \frac{h}{3}\left[y_0 + 4y_1 + 2y_2 + 4y_3 + y_4\right].$$

We can break the interval from a to b into more pairs of panels. For a total of n panels, with n an even number, the 1/3 rule becomes:

---

[6]Note that the quadratic shown for our example (Fig. 3.7, p. 62) overestimates the area under the true function, at least for the left-hand panel.

$$\int_a^b y \, dx \approx \frac{h}{3}\left[y_0 + 4y_1 + 2y_2 + 4y_3 + 2y_4 + \ldots + 2y_{n-2} + 4y_{n-1} + y_n\right].$$

Simpson also derived a 3/8 rule, which applies over *three* panels:

$$\int_{x_0}^{x_0+3h} f(x) \, dx \approx 3h\left[\frac{f(x_0) + 3f(x_1) + 3f(x_2) + f(x_3)}{8}\right].$$

Use the 3/8 rule only when necessary, because it is less accurate than the 1/3 rule. You really only need it when you want to integrate a function for which you have evenly spaced *tabulated* values over odd numbers of panels.

Simpson's rules are relatively easy to understand and use, and are adequate for many purposes. However, be aware that there are other more efficient numerical integration methods that are preferable if you have a lot of heavy-duty number-crunching to do. You will find some of these well described (including, for example the efficient *Romberg* method) in Press et al. (1992). In MATLAB, the *quadl* procedure is effective.

3.5 Differentiation-Integration Contrasts

Differentiation can be contrasted with integration in several interesting ways:

- Differentiation and integration are inverse processes relative to one another.

- Differentiation is straightforward, and integration is not.

- Differentiation is less "robust" than integration.

Let's consider these points in turn.

Differentiation and Integration Are Inverse Processes

This is stated formally by the "Fundamental Theorem of Calculus," which declares that if $f(x)$ is continuous over $a \leq x \leq b$ and if

$$\frac{dF}{dx} = f(x) \text{ when a} \leq x \leq b$$

then

$$\int_a^b f(x)\,dx = \int_a^b \frac{dF}{dx}\,dx = F(b) - F(a).$$

This is proved in many calculus texts.

In terms of indefinite integrals, this theorem tells us that if $dF/dx = f(x)$, then $\int f(x)\,dx = F(x) + C$. In words, the theorem essentially means that if a function f is the derivative of some other function F, then F is the integral of f.

Completing the graphical exercise at the end of this chapter may help you to gain an intuitive feeling for the inverse nature of the two processes.

Integration Is Not Always Straightforward

In contrast to the situation for differentiation (Chapter 4), there are many functions that cannot be integrated analytically so as to yield a result in closed form. For example, recall the function

$$f(x) \stackrel{\text{def}}{=} \frac{\sqrt{\cosh\{\sin[\log(x+1)]\}}}{(\sqrt{x}+1)^2}$$

from the previous chapter. Although we found the derivative of this function on p. 26, its integral over the range $1 \le x \le 3$ almost certainly doesn't exist in closed form. That is, probably no mathematician could find a function $F(x)$ with a finite number of terms such that

$$\int_1^3 f(x)\,dx = F(3) - F(1)$$

or such that

$$\int f(x)\,dx = F(x) + C.$$

If we ask MATLAB to find that integral using

```
syms x
f=sqrt(cosh(sin(log(x+1))))/(sqrt(x)+1)^2
int(f)
```

it returns an error message, along with

```
ans=int(cosh(sin(log(x+1)))^(1/2)/(x^(1/2)+1)^2,x),
```

which merely restates the problem, but provides no answer. However, you can estimate the value of the integral *numerically*, by Simpson's rule, for example. The value is approximately 0.419. To tell MATLAB to integrate that function numerically from 1 to 3, one choice would be to enter[7]

```
f = inline('(cosh(sin(log(x+1)))).^(1/2)./...
            (x.^(1/2)+1).^2')
intval = quadl(f,1,3)
```

MATLAB replies with `intval=0.4190`. The quadl function in MATLAB uses a process called Lobatto quadrature, which is more efficient and accurate, but more complicated, than Simpson's rule. Note that the definition of the function f requires dots (periods) in front of the *, /, and ^ operators (but not in front of + or -). This has to do with MATLAB's internal use of vectors while performing this integration.

Differentiation Is Not Robust

Integration is very *robust* to approximations and discontinuities, compared with differentiation, which is sensitive to such things. For example, suppose an ecologist measured the densities (in trees per acre) of balsam firs of various ages in a forested area, and found the age-density curve shown by the solid line in Fig. 3.8, p. 68 (J. Hett, personal communication, 1970).

The true curve suggests that there had been cycles of high and low reproduction in the past. A mathematical modeller interested in the overall decline of density with age might want to approximate the true solid curve with the dashed one. Note that the integrals under the two curves would be very similar (because integration averages out fluctuations), but S2, the derivative of the approximate curve, can be very different from S1, the derivative of the true curve, at many points.

---

[7]In MATLAB, an ellipsis (...) at the end of a line normally indicates that the command is continued on the next line. However, you can't break a string (material between quotes) that way, so you'll have to enter this as one line to run it.

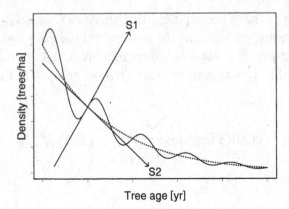

Tree age [yr]

Figure 3.8: The age-density relationship for balsam fir in a region of New Brunswick. Solid line—true curve; dashed line—an approximate curve. Note that the two curves have very similar integrals (areas under the curves), but that the slopes of the relationship for trees of certain ages differ drastically between the true and approximate curves.

3.6 Exercises

1. Review of basic integration: Integrate the following functions by substitution, with the help of the tables on pp. 60 and 131 or by any other appropriate method. You may also want to try integrating them using MATLAB.

$$\text{A.} \int \frac{\log x}{x}\, dx; \quad \text{B.} \int \frac{x}{x^2+1}\, dx; \quad \text{C.} \int_0^{\log 2} e^t\, dt$$

$$\text{D.} \int (e^t + e^{-t})^2\, dt; \quad \text{E.} \int_0^1 t^2 e^t\, dt; \quad \text{F.} \int_0^2 x e^{-x}\, dx$$

2. Using analytic integration, find the value of the integral

$$I = \int_1^{2.5} (x^2 - \sqrt{x})\, dx$$

3. Check your result in the previous exercise using Simpson's 1/3 rule with six panels.

4. Find the mean value of the function $y = 3e^{-0.5} - 2$ analytically, over the range $a \le x \le b$. Work symbolically as long as you can, then plug in numerical values to obtain the final numerical value (for $a = 1$, $b = 3$) only at the very end.

5. Consider a 12-m wide stream that flows north to south, and suppose that its depth varies with distance x from the western shore approximately as $d = ax + bx^2 + cx^3$ for $0 \le x \le 6$ m. Assume the stream bed has the reverse shape from $6 \le x \le 12$ m. What is the volume of water in a slice of the stream one meter along its length?

6. (This one's a little harder.) Consider a circular farm pond of radius 8 m, in which the depth varies with distance x from the shore according to $d = a + bx + cx^2$, for $0 \le x \le 8$ m. What is the total volume of water in the pond?

7. The human population was approximately 0.5 "gigabod" at 1.5 millenia C.E. (common era) and 3 gigabod at 1.964 millenia C.E. These two data points can be used to determine uniquely (but probably not accurately) the two constants in each of the following models for population growth:

Linear: $N_L(t) = A + Bt$ (1)
Exponential: $N_e(t) = ae^{bt}$ (2)

Based on each of these models, estimate:

A. the mean population over the whole time interval between the two times.

B. the rate of population increase at time 1.732.

8. Consider a population that grows logistically, according to this equation, modified from Eqn. 1.6, p. 9:

$$N(t) = \frac{Ke^{rt}}{e^{rt} + \dfrac{K - N_0}{N_0}}.$$

Use Simpson's one-third rule with four panels to estimate the mean population over the time period from $t = 0$ to $t = 2$ years, when the parameters are $K = 1000$ animals, $r = 0.03$ yr^{-1}, and $N_0 = 7$ animals.

9. You work for a city that owns a small reservoir that is a minor part of its water supply. Three flows—via an aqueduct entering, a stream entering, and an aqueduct leaving the lake—are measured

on an instantaneous basis at 7 a.m. (0700 hours) and 7 p.m. (1900 hours) every day. These flows are used to estimate changes in the volume of water stored in the reservoir, under the assumption that the three flows all change fairly smoothly. (That assumption sometimes breaks down during heavy storms, but no storms occur during the period you will work with here.)

Based on past calculations, we know that the reservoir contains about 7.5×10^6 m^3 of water at 0700 on the first day shown. Given the flows listed in the table below, use numerical integration to estimate the amount of water stored at 0700 on each of the next two days. Be sure to check units, and to show your unit check(s).

Day	Time	$Q_{\text{duct in}}$ [m^3 s^{-1}]	$Q_{\text{stream in}}$ [m^3 s^{-1}]	$Q_{\text{duct out}}$ [m^3 s^{-1}]	Net in [m^3 s^{-1}]
1	0700	3.85	0.97	4.25	0.57
1	1900	3.78	1.03	4.31	0.50
2	0700	3.75	1.04	4.36	0.43
2	1900	3.81	1.08	4.40	0.49
3	0700	3.86	1.04	4.43	0.47

10. Most of the water provided to NY City by the NYC DEP is unfiltered, because the source waters (mainly in the Catskill Mountains) are remarkably clean. As one condition for avoiding having to build a filtration system (at a cost of many billions of dollars), the US EPA requires the city to keep concentrations of *Escherichia coli* and other indicators of fecal contamination at specified low levels.

As an aid in meeting that condition, DEP hydrologists study the loading of FCBs (fecal coliform bacteria) discharged by streams into the City's reservoirs. Here we consider data obtained for Suburban Creek, which drains a largely residential area north of White Plains into the Kensico Reservoir. Potential sources that might add FCBs to that creek are pet and wildlife feces, leaky sewers, and malfunctioning septic systems.

The hydrologists obtain data for hourly discharge q (flow of water) in m^3 hr^{-1} of water, as it varies with time through a year. They also take water samples once every four hours for a full year[8], and send

---

[8]In reality, this would be prohibitively expensive. They actually sample intensively only during storm events

these to the lab for determination of FCB concentrations c, which we will express[9] in CFU m^{-3}.

Suppose we can express q and c as known, continuous functions, say $q = f(t)$ and $c = g(t)$ for one year of interest. (This could be done approximately using mathematical structures known as *cubic splines* for example.) We now wish to estimate the total number of FCB CFUs discharged into Kensico via this stream in the year under study.

One of the team, John Doe, suggests calculating the mean hourly water discharge $\bar{q}$ for the year and the mean FCB concentration $\bar{c}$ over the year. Then, he would estimate total FCB loading T [the number of FCB CFUs over the year] as $T = \bar{q} \times \bar{c} \times N$, where N is the number of hours in the year. Another of the team points out that the FCBs can sometimes be flushed out of the stream's watershed early in a storm, so that maximum c values may often occur with small q values, and conversely.

Your task is to describe an analysis that would take that latter phenomenon into account and would provide a more accurate estimate of the year's loading of FCB into Kensico (from this stream) than Doe's calculation could provide. Make your description as mathematically explicit as you can, given that you don't yet know the specific forms of q and c. Be sure to

- Describe your method and reasoning in full sentences,
- Draw any hypothetical sketches that will help to illustrate your reasoning,
- Analyze units, and
- Describe why your proposed method should work better than J. Doe's method.

Hint: It may help you to think back to birds colliding with power lines. It may also help you to think about hourly FCB loadings for a few "representative" hours during the year.

11. The data in the table below represent *instantaneous* flow rates Q, phosphorus concentrations C, and phosphorus loading rates R ($=$

---

[9]CFU=colony-forming units

$QC/1000$) entering a constructed wetland. Use Simpson's rule to estimate the total loading of P, expressed in *kilograms*, into the wetland over the six-month period. To simplify the calculations, assume that each month comprises exactly 30 days.

State your result in a sentence, and supply a unit check, including a statement about whether the units do check or do not.

Date	Q $\left[\text{m}^3 \text{ da}^{-1}\right]$	C $\left[\text{mg m}^{-3}\right]$	R $\left[\text{g da}^{-1}\right]$
Apr 1	2480	39.6	98.21
May 1	1932	44.2	85.39
Jun 1	1756	48.0	84.29
Jul 1	1288	48.2	62.08
Aug 1	945	47.8	45.17
Sep 1	1120	43.6	48.83
Oct 1	1432	40.6	58.14

12. Toxicologists who study acute responses to inhaled substances sometimes make use of "Haber's Law," which is really just a first approximation rather than a scientific "law" (whatever that means). This principle states that the physiological response to an inhaled toxin will tend to be about the same when the *product* of concentration C inhaled times the duration of exposure T is held constant. Thus, a rat breathing air containing 1 mg m^{-3} of formaldehyde for two hours would be expected to respond similarly to a rat breathing 2 mg m^{-3} for one hour. Let us denote the effective total exposure as E, where $E = CT$. If C and T have the units given above, what are the units of E?

For exposures to *varying* concentrations, E is obtained by integrating (accumulating) the concentration over the exposure time. Suppose that you spill some volatile substance in a lab, and the concentration carried by the air ducts to a nearby classroom varies as

$$C(t) = ae^{-b(t-k)^2} \quad \text{for } 0 \le t \le 2k,$$

$$C(t) = 0 \text{ otherwise.}$$

Then

A. Obtain the integrated concentration over the $2k$ minutes that this episode lasts. Because this function can't be integrated analytically, estimate its value numerically for the case when $a = 2$ mg m^{-3}, $b = 0.02$ min^{-2}, and $k = 20$ min. Use Simpson's rule with 8 panels. Be sure to show your work in detail.

B. Explain briefly how you might decide whether eight panels yield a sufficiently accurate answer. (Just explain what you would do— you don't have to carry out the procedure.)

C. What is the *average* concentration over the 40-minute period of interest here? (Express your answer symbolically, and then as a numerical value.)

D. Is it true that the average, when multiplied by the 40-minute duration, is the same as the integrated E value?

13. There are several kinds of situations that call for numerical integration, such as:

- having a function in "formula" form for which no analytic integral can be found.

- checking an analytic integral, when one *has* been found.

- having a functional relationship (e.g., some response y that varies with time) that is known only as a table of measurements, and not as a formula.

The problems above have dealt with the first two of these situations. The present problem provides an example of the third.

Convective heat loss between an object and the air surrounding it is often modelled with a *convection coefficient* h_c, such that the energy loss from the object is given by $q = h_c A (T - T_A)$, where h_c has units like cal cm^{-2} min^{-1} deg^{-1}, A is the object's area in cm^2, T is the object's temperature, and T_A is the temperature of the air (or other fluid) surrounding the object. The convection coefficient, in turn, can be expected to be proportional to the square root of the object's length in the direction of the wind, when the wind flows along in the plane of the object. When the length of the object varies, then h_c could be calculated from the mean square root of the length.

Figure 3.9: An aspen leaf with lines indicating locations where the measurements in Table 3.2 were taken. The lines are 0.16 cm apart.

1.27	3.69	3.29	1.47
2.09	3.75	3.10	1.16
2.62	3.78	2.90	0.63
2.98	3.77	2.68	0.35
3.22	3.72	2.51	0.11
3.43	3.62	2.22	
3.60	3.48	1.92	

Table 3.2: Measurements [cm] of the lengths of the lines shown in Fig. 3.9. (Read numbers top to bottom in the first column, then in the second column, etc.)

The aspen (*Populus tremuloides*) leaf shown in Fig. 3.9, has the lengths (in cm) provided in Table 3.2, measured along the lines shown in the figure. Use these data to determine the mean square root of width, using Simpson's rule. If the number of panels is not an even number, then use the 3/8 rule once, somewhere in the part of the range where the widths vary the least.

Also, describe how you could use Simpson's rule to estimate the area of this leaf. (You do not have to carry out these calculations.)

14. Decay of leaf litter on the forest floor is of interest for many reasons, not the least of which is that we would all soon be buried alive if dead things didn't rot. In addition, the decay process releases nutrients back to the soil, and is part of the global carbon cycle. Leaf litter contains many components, but two important ones are cellulose and lignin—cellulose is fairly easily decomposed by microorganisms like fungi and bacteria, while lignin is more re-

calcitrant. One model that accounts for "slow" and "fast" components of litter is

$$m = f_0 \exp(-k_f t) + s_0 \exp(-k_s t) = f_0 e^{-k_f t} + s_0 e^{-k_s t},$$

where

$m =$ total litter mass remaining in a sample [g]
$f_0 =$ initial fast mass [g]
$s_0 =$ initial slow mass [g]
$k_f =$ decay constant for the fast mass [da^{-1}]
$k_s =$ decay constant for the slow mass [da^{-1}]
$t =$ time after leaf fall [da].

(All these values are positive at all times.)

Assuming that relationship, and that no new litter falls during the time of interest, determine the mean amount of litter mass present between any two times, t_1 and t_2 within that decay process. Provide a complete unit check.

15. Make an enlarged photocopy of Fig. 3.10 on p. 76 if you can, and for each of the $f(x)$ functions there, sketch the first and second derivatives in the top two rows of panels. In the bottom row, sketch at each x the integral of the function from zero to that value of x. Remember that integrals accumulate. Your goal should be to sketch the general tendencies correctly—you need not be quantitatively correct.

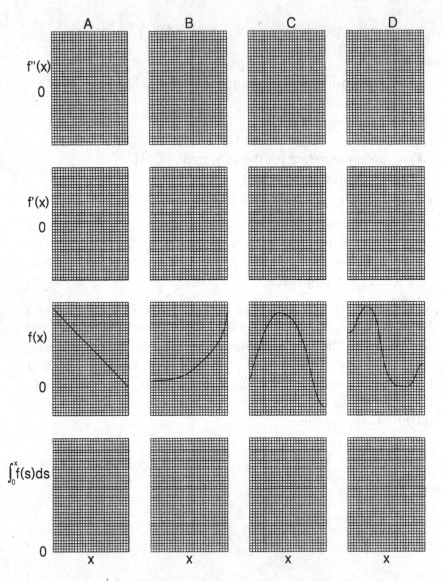

Figure 3.10: Graphical exercise for understanding the basic meaning of derivatives and integrals.

3.7 Questions and Answers

1. Why is $\log(x)$ the integral of $1/x$?

   • The following doesn't explain *why*, but it helps to confirm the relationship. The Taylor series for $\log(x)$ at $a = 1$ happens to be

   $$(x - 1) - \frac{1}{2}(x - 1)^2 + \frac{1}{3}(x - 1)^3 - \frac{1}{4}(x - 1)^4 + \frac{1}{5}(x - 1)^5 + \dots.$$

   If you differentiate that term by term, you get the Taylor series for $1/x$ at $a = 1$, which is $1 - (x - 1) + (x - 1)^2 - (x - 1)^3 + (x - 1)^4 - (x - 1)^5 + \dots$.

2. Please provide an example of calculating an integral, like the one in the class exercise today.

   • Okay. The function was $0.1(1 + x + x^2)$. Let's call its integral from 0 to 10 "I." Using the rule that the integral of a constant times a function is the constant times the integral of the function, we get $I = 0.1 \int (1 + x + x^2) dx$. Next use the rule that the integral of a sum is the sum of the integrals to get

   $$I = 0.1 \left[\int 1 \, dx + \int x \, dx + \int x^2 \, dx \right].$$

   The first of those integrals (from 0 to 10) gives 10-0. The second gives $(10^2 - 0^2)/2 = 50$. The third gives $(10^3 - 0^3)/3 = 1000/3$. Finally, then, $I = 0.1(10 + 50 + 333.\overline{3}) = 39.\overline{3}$. To get the average of the function over some range, you divide by the range (10-1), so the average value was $3.9\overline{3}$.

3. There were several questions about Simpson's rule, so let me try to answer them all at once.

   • In various numerical methods, we often use polynomials in one way or another. Even if we don't use Taylor series explicitly (we don't with Simpson's rule), our brief look at them tells us two important things:

   • All other things being equal, polynomials with more terms provide better approximations to general functions than polynomials with fewer terms. (This is especially true for narrow ranges of the x variable.)

• Even low-degree polynomials often provide good approximations if we work within a narrow enough x range.

There are lots of ways one *could* approximate functions to estimate integrals. There are methods, e.g., "Gaussian quadrature," that involve fitting a single 16th degree polynomial (for example) to the function being integrated. Alternatively, one can use relatively low degree polynomials, and break the range of integration into lots of narrow subranges. That's what we do when we use Simpson's rule.

Simpson's rule is particularly effective because, even though we derived it to approximate a function with a quadratic, it actually provides exact integrals for cubics. That means that it approximates more complicated functions with cubic approximations, too.

4. How do computers integrate functions (with references to some related questions)?

• Most of the time they don't use Taylor series explicitly, unless you tell them to. Programs like Maple (and MATLAB symbolics) and Mathematica (and some others) "know" how to use the general rules of symbolic integration (including integration by parts, for example), and can integrate many functions that have analytic integrals. For example, if I want to evaluate $\int \sin(x) \exp(-bx) \, dx$, I enter

```
syms b x
f=sin(x)*exp(-b*x)
simplify(int(f))
```

in MATLAB, and it yields (without the constant of integration, remember)

```
ans = -exp(-b*x)*(cos(x)+b*sin(x))/(b^2+1).
```

(The "simplify()" is not necessary, but often gives a cleaner looking answer.) This is MATLAB's notation for

$$- \exp(-bx)\frac{\cos(x) + b\sin(x)}{b^2 + 1}.$$

MATLAB can also do *numerical* integration, as was shown on p. 67.

With any programs like MATLAB, keep in mind the following warning (about the similar Maple) from Kofler (1997): "... Therefore you should always look at Maple results with a certain degree of scepticism. The better your knowledge of the underlying mathematics, the higher the probability of detecting such errors and possibly finding an alternative formulation for which Maple supplies the correct result...."

5. Why can't you use Simpson's 1/3 rule with an even number of data values (if all you have is a table of data, not a formula)?

   • An even number of values corresponds with an odd number of panels. But Simpson's 1/3 rule always works with sets of two panels at a time, as with $(y_1 + 4y_c + y_2)$. Simpson's 3/8 rule is not as accurate, but it allows you to work with three panels.

6. What was the 'handy trick' you mentioned to check if you've used Simpson's rule correctly?

   • Actually, I was suggesting the opposite—using Simpson's rule to check whether you have done some symbolic integration correctly. If you integrate some messy function (by substitution, or by parts, or with tables, or even with software), it's not a bad idea to check your result. One way to check is to differentiate, and see whether you get your original function back, but sometimes the resulting algebraic forms can look so different that it's hard to tell. Another way to check (and you'll see this in some of this chapter's exercises) is to see whether Simpson's rule gives you essentially the same numerical result as your analytic integral, over some specific range.

7. Is Simpson's rule useful for real applications, such as engineering work?

   • Yes, it is a fairly standard way of performing numerical integration. In particular, it's pretty easy to work with Simpson's rule in a spreadsheet if you don't have (or don't want to use) fancy software like MATLAB, and it's accurate enough for many purposes. Simpson's rule is about the best you can do if you have only a table of measured values to represent a continuously varying quantity, too. A lot of numerical integration these days is performed using more complicated methods, but those are not at all easy to implement in a spreadsheet. (Software like MathCad and MATLAB, and calculators that have integrators built in probably use more sophisticated, but more complex, methods. When I worked for the NY City water supply, one of my staff used Simpson's rule to integrate flows into and out of the City's reservoirs.

8. If you have a set of measured data taken at uneven intervals, how would you use Simpson's rule?

• You can't. If I had that situation, and needed to estimate the integral under the curve that the data represent, I would probably use repeated applications of the trapezoidal rule between each set of two numbers. Another approach would be to use approximating functions called cubic splines, and to integrate those.

9. How can the $p(z)$ expression you wrote for the 'bell curve' be an equation for that function, since you said there is no algebraic way to write it?

• The $p(z)$ equation (the integrand, or the lower-case $p(z)$ function inside the integral in Eqn. 3.7 in the notes) is the equation for the bell itself. What I said was that the *integral* of that curve, the $P(z)$ (capital P) that represents the actual probabilities of being between any two numbers, doesn't have an algebraic representation. That is, although the function $x^2 + 2$ has the integral $(x^3/3) + 2x + C$, you can't integrate $p(z)$ to get a symbolic form anything like that. In statistics, that $p(z)$ is called a "probability *density* function;" it does not represent actual probabilities.

10. Please explain again why we divide h by 3 in Simpson's rule.

• I'll explain at two levels. The first is that the "3" came out of the derivation, and it has to be there to make the answer exactly correct when you integrate a function that happens to be a quadratic. The second is (and this may be what you are asking for) that Simpson's rule can be written in the form of the *width* of the area of interest $(2h)$ multiplied by the average *height* of the 'y' variable over that range. That average height is a "weighted average," namely $(y_1 + 4y_c + y_2)/6$. The 2 multiplied by (1/6) leaves you with 1/3. In fact, the rule we learned is sometimes called "Simpson's one-third rule." There is also a three-eighths rule that is useful for limited special cases.

11. What exactly does Simpson's rule tell you? Is it the same as taking a definite integral?

• It is similar in intent. Specifically, it is a numerical method for approximating the value of a definite integral, for use primarily when you can't obtain that value analytically.

12. If you have a function that can't be integrated analytically, could you represent it graphically, and then approximate the integral with infinitesimally small Δx's?

• Well, if you make them infinitesimally small, literally, then you would need an infinite number of them, so that process would take you a very long time <grin>. In the days BC (before computers), I've heard of people drawing a curve, cutting out the area, weighing it, weighing a known rectangular area, and using that information to estimate the unknown area. I've also used an area-measuring device called a planimeter used. But it's easier to use one of the numerical integration routines we will discuss, and that might be "good enough for government work" (as the saying goes). The aspen leaf exercise demonstrates that process.

13. Your example in Fig. 3.5 was very extreme. How well does the Weather Bureau's averaging method work for realistic cases?

• When I was an undergraduate in Boulder, Colorado, the temperature dropped one day from something like 90 to 30 F in about six hours. Still, I agree that kind of thing doesn't happen very often. My example was meant more to illustrate what the mean of a continuous function is, rather than to criticize the WB's method. Given all the uncertainties in temperature measurements, their method is probably good enough for most purposes. There are places where sea breezes cool things off rapidly in the evening, day after day, in the summer. If that were a regular pattern, the simplistic [max+min]/2 could be biased by an amount that might matter for some purposes.

14. Isn't it easier to change the limits of integration to limits for the new variable when you make a substitution, rather than doing it the way you did?

• You can do it whichever way seems easier to you. It's a matter of personal preference—either works if you're careful.

Chapter 4

Ordinary Differential Equations

4.1 How ODEs Arise

Although derivatives are important in environmental science in ways described in Chapter 2, their greatest importance may be as components of differential equations. Until now, we have mostly translated story problems into algebraic or transcendental[1] equations involving specific variables directly. That is, we might be able to work out that some response y is a given function of some influencing variable x. Often, though, we don't have knowledge of that direct kind. Instead, we have two other kinds of knowledge:

- Given any value x of the causative (independent) variable and any value y of the response (dependent) variable, the *rate of change* of y with x (that is, dy/dx) is a known function of x and y. In the environmental sciences, the independent variable (x) is often time, so x is often replaced by t in these notes.

- At x_0 or t_0 (some particular value of x or t), the value of y is y_0.

Situations like these lead to differential equations, which describe the *rates* at which state variables like mass, concentration, temperature,

---

[1]Technically, algebraic equations are those involving polynomials only. Logarithms, exponentials, and trigonometric functions are known as transcendental functions. We won't pay further attention to this distinction, however.

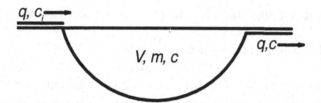

Figure 4.1: Definitions of quantities in the mercury-in-lake problem. q=flow into and out of the lake [m^3 da^{-1}]; c_i=concentration of Hg in the water flowing in [mg m^{-3}]; $m = m(t)$=mass of Hg in the lake water [mg] as it varies with time; V=volume of water in the lake [m^3], a constant here; $c(= m(t)/V)$=concentration of Hg in well-mixed lake water [mg m^{-3}].

pressure, population size, or other conditions change with time, with distance along some spatial coordinate, or occasionally with some other variable. Our ultimate interest is often in the states rather than in the rates, and we obtain information about the states by solving the differential equations (pp. 89ff). The reader will find that careful attention to the distinction between rates and states will help in learning to understand and use differential equations effectively.

We begin our study of these valuable mathematical tools with three examples. We will first *derive* the equations (write the mathematical statements for the processes being modelled), and later solve them.

Example 1

We return to the mercury-in-lake problem posed earlier (p. 12), for the lake diagrammed in Fig. 4.1. Let's simplify matters with some "heroic assumptions"—that all mercury is dissolved in the water and none of it evaporates, drops to the bottom with sediments, is taken up by organisms, or is involved in chemical reactions. We'll also assume that flow of water into the lake equals flow out, and both are constant (so lake volume doesn't change with time), and that c_i remains constant through time.

Let $m = m_0$ (the *initial condition*) at some starting time $t = 0$, where m is the total mass [mg] of mercury in the lake. In the future, for *any* time period of length Δt (not only at the first time step!), a simple mass balance tells us that $m(t + \Delta t) = m(t) + (m$ added in time $\Delta t) - (m$ lost in time $\Delta t)$.

If Δt is short enough, then for the processes shown above

$$m(t + \Delta t) \approx m(t) + c_i q \Delta t - \frac{m}{V} q \Delta t. \qquad (4.1)$$

Note the use of "$\approx$" and not "$=$", which will be explained shortly. Here, c_i is the (assumed constant) mercury concentration [mg m^{-3}] in the incoming stream, q is the discharge of water [m^3 da^{-1}] of both the inlet stream and the outlet stream, and V is the volume of water [m^3] in the lake. The units of the various terms in the equation above are thus

$$\text{mg} = \text{mg} + \frac{\text{mg}}{\text{m}^3} \frac{\text{m}^3}{\text{da}} \text{ da} + \frac{\text{mg}}{\text{m}^3} \frac{\text{m}^3}{\text{da}} \text{ da}.$$

With a little algebra, Eqn. 4.1 becomes

$$\frac{m(t + \Delta t) - m(t)}{\Delta t} \approx q\left(c_i - \frac{m}{V}\right).$$

This form is only approximate because the mass m, and hence the loss rate of mercury from the lake, are both changing continuously—those variables are not constant for the whole period from t to $t + \Delta t$.

We can save the day by taking the limit as Δt goes to zero, to yield

$$\lim_{\Delta t \to 0} \frac{m(t + \Delta t) - m(t)}{\Delta t} = \frac{dm}{dt} = q\left(c_i - \frac{m}{V}\right),$$

with the *initial condition* that $m(0) = m_0$. This is a differential equation, in this case a rule for the rate at which mercury mass m will change at any given time, based on what m is present at that given time. Of course, we really want to know how m changes with time; i.e., $m = f(t)$. That would be the *solution* to the DE (differential equation), and we will learn shortly how to obtain it. To that end, note for now that the DE could be written as $m' = a + bm$, with $a = qc_i$ and $b = -q/V$, where a and b are both constants.

Example 2

We have here the tale of an oddly-motivated swallow headed for Capistrano one spring. (The model is fanciful, but it helps to illustrate some important concepts.) At any instant, she flies toward her goal at an airspeed that is one-fourth of her distance D [km] from

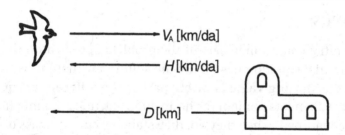

Figure 4.2: A swallow flying toward Capistrano. V_A is the bird's variable airspeed, relative to the wind [km da^{-1}], H is the constant wind speed [km da^{-1}], and d is her (variable) distance from the mission [km].

the mission per day. This airspeed is by definition relative to the air, not the ground; further, it is an *instantaneous* rate that can change from instant to instant. In other words, as she gets closer to her goal, she slows down. Unfortunately, her airspeed is also relative to a headwind of H km da^{-1}.

If we knew her distance D_0 at some starting time t_0, it would be convenient to be able to write down her $D(t)$ function for future times, so the tourists at Capistrano would know when to expect her. But we don't really have that knowledge—all we know about is *rates* of motion. So again, we derive a differential equation, which for this problem we can write directly. We have

$$\frac{dD}{dt} = \text{(Distance increase rate)} - \text{(Distance decrease rate)}$$

$$= H - \frac{D(t)}{4}, \tag{4.2}$$

with the initial condition that $D(t_0) = D_0$. Units are

$$\frac{km}{da} = \frac{km}{da} - \frac{km}{da}.$$

This is of the form $D' = a + bD'$ (with $a = H$ and $b = -1/4$), or more generally of the form $y' = a + by$.

Be sure to note that this equation applies at all times into the future, not just at the starting moment. A quick check shows that dD/dt, H, and $D/4$ all have units of km/da, so our units are at least consistent. Finally, note the initial condition (IC)—we will need both the DE and the IC to obtain a solution, which again we will take up shortly. At this point, it is worthwhile to sound several warnings about confusions that students sometimes experience.

WARNINGS!

- The initial condition is part of the problem specification, and the differential equation has no unique solution without it. Nevertheless, **the starting value is rarely part of the differential equation itself**. (Chemists sometimes do include starting amounts in their rate equations—when they do, they state the current mass of some reactant as the starting mass less what has been lost to the reaction so far.)

- To be more specific, one should **not** write Eqn. 4.2 as

$$\frac{dD}{dt} = H - \frac{D_0}{4}.$$

This form would be correct for just an instant at $t = 0$, but after that (after the swallow's distance from her goal changes from D_0) it would overestimate her airspeed as she gets closer to her target and slows down.

- Also, **do not** write the DE as

$$\frac{dD}{dt} = D_0 + H - \frac{D}{4}.$$

This form makes no sense at all, since the distance D_0 is not a rate and even has different units from the other three terms. It is D, and not its derivative, that starts at D_0. The *rate of change of D* may or may not ever have the same numerical value as D_0, but in any event that would be irrelevant, since distance D and its rate of change are different quantities with different units, and so can't be added.

This may suggest the greatest difficulty beginners have with DEs. It is important to distinguish carefully between the state or condition of a variable on the one hand, and the rates at which it changes on the other.

- I noted above that $D/4$ was the swallow's *instantaneous* rate of flight (relative to the wind) toward the mission. This means that at the particular moment when she is 100 km away, her forward airspeed would be 25 km/da (or about 1.042 km/hr, or 17.36 m/min, or 28.94 cm/sec). But this *doesn't* mean that she will actually travel 25 km in the next 24 hours, nor even 28.94 cm in the next second. The reason is that as she gets closer to her goal, even just a little bit, she slows down. So her speed at $D = 100$ km is her speed for only an instant. This is a good example of why calculus and the concept of the limit are needed.

Suggested Exercise:

The rates in the swallow problem were sufficiently simple that we could write down the overall differential equation directly. But you might find it instructive also to use the general format of

$$D(t + \Delta t) \approx D(t) + \Delta t (D \text{ increasing rates} - D \text{ decreasing rates})$$

and then to apply an appropriate limiting process to obtain that same DE.

Example 3

In Fig. 4.3, a point source adds solids to a river at a position $x = x_0$. We assume the river water is generally well mixed, and that the new solids mix instantly with the water from upstream, yielding a solids concentration of c_0 g m^{-3} at x_0. The density of the solids is such that they sediment out to the stream bottom; 5% of those present in any small volume drop out per hour. (Actually that is the instantaneous

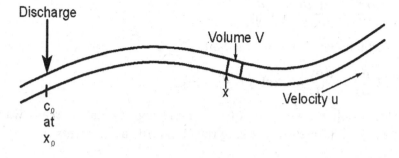

Figure 4.3: Variable definitions for Exercise 3. Here x [m] is the distance downstream from the discharge point, x_0; c_0 [g m^{-3}] is the sediment concentration at $x = x_0 = 0$; $c(x)$ is the concentration at any distance x; V [m^3] is the volume of water within a short reach of stream, and u [m s^{-1}] is the stream velocity.

sedimentation rate, so the net loss would be somewhat less than 5% per hour.)

Our problem is to determine the concentration $c(x)$ of solids in the water at different distances downstream. Again we have the situation of knowing the value of the *state variable c* at one particular point, and the *rate* at which c would decrease from any particular value. That knowledge will ultimately prove sufficient to give us $c(x)$ for all $x > 0$, but we will have to work to obtain that result.

Our approach is to consider a small volume V of water (sometimes called a *control volume*) at some distance x and corresponding time t downstream from the discharge. If we let m be the mass in grams of solids present in V, then

$$\frac{dm}{dt} = -km = -kcV, \tag{4.3}$$

where $k = 5\%$ per hour, or 0.05 hr^{-1}. **(Warning: Remember always to convert percentages to fractions in equations like this.)** Note that $dm/dt = [\text{g hr}^{-1}]$ and $kcV = [\text{hr}^{-1}][\text{g m}^{-3}][\text{m}^3] = [\text{g hr}^{-1}]$ also.

Our goal here is to find the concentration $c = f(t)$, or better yet, $c = f(x)$. Of course, concentration is mass per volume, so $c = m/V$ and $m = Vc$. From this,

$$\frac{dc}{dm} = \frac{1}{V} \text{ or } dm = Vdc \text{ or } \frac{dm}{dc} = V.$$

Using the chain rule (p. 25), we change the LHS of Eqn. 4.3 to

$$\frac{dm}{dt} = \frac{dm}{dc}\frac{dc}{dt} = V\frac{dc}{dt}$$

from which (because V is constant)

$$V\frac{dc}{dt} = \frac{dm}{dt} = -kcV \text{ or } \frac{dc}{dt} = -kc.$$

Our problem was to find $c(x)$, which suggests that we really want an expression for dc/dx. Using the chain rule again yields

$$\frac{dc}{dx} = \frac{dc}{dt}\frac{dt}{dx}.$$

We know dc/dt, but what is dt/dx? Its dimensions are time/distance, or the inverse of velocity. In fact, if the stream velocity is u [m hr^{-1}], then

$$u = \frac{dx}{dt}, \text{ from which } \frac{dt}{dx} = \frac{1}{u}.$$

Putting all this together yields

$$\frac{dc}{dx} = \frac{dc}{dt}\frac{dt}{dx} = -kc\frac{1}{u} = -\frac{k}{u}c,$$

which again is of the form $y' = a + by$, with $a = 0$. The initial condition is of the form $c(0) = c_0$.

4.2 Solution of Simple First-Order ODEs

We have now *derived* three relatively simple first-order ordinary differential equations:

$$\frac{dm}{dt} = q\left(c_i - \frac{m}{V}\right); \quad m(0) = m_0 \quad \text{(mercury)}$$

$$\frac{dD}{dt} = H - \frac{D}{4}; \quad D(0) = D_0 \quad \text{(swallow)}$$

$$\frac{dc}{dx} = -\frac{k}{u}c; \quad c(0) = c_0 \quad \text{(solids in stream)}.$$

What we want to know are $m(t)$, $D(t)$, and $c(x)$, but all we have is equations for the rates of change of these variables. So, we now *solve* the differential equations to obtain the desired equations for m, D, and c. This process is sometimes referred to as *integrating differential equations*.

Our technique of working in symbolic, general terms now becomes particularly useful, because all three of these DEs turn out to have the same general form, *i.e.*:

$$\frac{dy}{dt} = a + by, \text{ with } y(0) = y_0.$$

That is, in all three rate equations, the rates of change of the state variables represented by y are linear functions of the dependent (state) variables. So, if we can solve this general form, we will have solutions for all three problems, and, it happens, for many other important differential equations that arise in environmental science.

Before taking on $y' = a + by$, let's solve four simpler cases to illustrate the general process. First, suppose $a = b = 0$, so

$$\frac{dy}{dt} = 0, \; y(0) = y_0.$$

If we multiply through by dt and integrate both sides using definite integrals, we obtain

$$dy = 0dt$$

$$\int_{y_0}^{y} dy = \int_{0}^{t} 0 \, dt$$

$$y - y_0 = 0 - 0$$

$$y = y_0 \text{ (for all t)}.$$

This solution should be obvious because if $y = y_0$ at one value of t, and if y never changes (as is implied by $y' = 0$), then y must equal y_0 at all times.

   The solution method above used *definite* integrals, with the initial condition providing the lower limits of the two integrals. A second approach uses *indefinite* integrals, with the constant of integration obtained from the initial condition. Let

$$dy = 0dt$$

$$\int dy = \int 0 \, dt$$

$$y + C_1 = 0 + C_2.$$

Now let $C \stackrel{\text{def}}{=} C_2 - C_1$ to obtain

$$y = 0 + C,$$

where C is the constant of integration that is required whenever one evaluates an indefinite integral. In this approach, we use the initial condition to solve for the numerical value of C. Because $y = 0 + C$ everywhere, and $y = y_0$ at $t = 0$, it is clear that C must equal y_0, so $y = y_0$ for all t.

   Next, let's solve the equation

$$\frac{dy}{dt} = a, \text{ with } y(0) = y_0, \text{ and } a \text{ a constant.}$$

This could represent the mass y of some material in a pond if the rate of accumulation a were constant. In this case we have

$$\int_{y_0}^{y} dy = \int_{0}^{t} a \, dt$$

from which $y - y_0 = a(t - 0)$, or $y = y_0 + at$. This is consistent with our prior knowledge that functions with constant derivatives are linear. Solving this equation using indefinite integrals works this way:

$$dy = a \, dt$$

$$\int dy = \int a \, dt$$

$$y = at + C,$$

where C is the constant of integration. Since $y_0 = a \cdot 0 + C$, we see that $C = y_0$, so $y = y_0 + at$, as before.

Next, consider

$$\frac{dy}{dt} = y, \ y(0) = y_0,$$

which says that the rate at which a variable y changes with time is equal at any time to its current value. Solving this by separation of variables takes the form

$$\frac{dy}{y} = dt$$

$$\int_{y_0}^{y} \frac{dy}{y} = \int_{0}^{t} dt$$

$$\log y - \log y_0 = (t - 0).$$

Because a difference in logs is the log of the quotient (p. 289), this becomes

$$\log \frac{y}{y_0} = t$$

$$\frac{y}{y_0} = e^t$$

$$y = y_0 e^t.$$

Note that this is consistent with the fact that $f(t) = \exp(t)$ is the only function we know that is equal to its own derivative!

The equation

$$\frac{dy}{dt} = by, \ y(0) = y_0$$

is only slightly harder. This equation represents a situation where the variable y increases at a rate proportional at any moment to its current size—accumulation of principal at a constant interest rate is an example. (Note that this is also mathematically the same equation as the simple population growth equation, $dN/dt = rN$.) Here, we "separate variables," dividing both sides by y and multiplying both by dt. Then we integrate:

$$\frac{dy}{y} = b \, dt$$

$$\int_{y_0}^{y} \frac{dy}{y} = \int_{0}^{t} b \, dt$$

$$\log y - \log y_0 = b(t - 0).$$

Because a difference in logs is the log of the quotient (p. 289), this becomes

$$\log \frac{y}{y_0} = bt$$

$$\frac{y}{y_0} = e^{bt}$$

$$y = y_0 e^{bt}.$$

Solving this equation with indefinite integrals is left for you as an exercise.

This brings us, finally, to our general equation,

$$\frac{dy}{dt} = a + by, \tag{4.4}$$

from which

$$\int dy = \int (a + by) \, dt.$$

Unfortunately this can't be solved directly because the y on the right-hand side is an unknown function of t. However, we can again separate variables (y on one side, t on the other) to obtain

$$\frac{dy}{a + by} = dt,$$

and integrate both sides:

$$\int_{y_0}^{y} \frac{dy}{a + by} = \int_{0}^{t} dt.$$

Here a simple substitution, $u \stackrel{\text{def}}{=} a + by$, helps. Then, because $du = b\,dy$, we have

$$\frac{1}{b} \int_{y=y_0}^{y=y} \frac{du}{u} = \int_{0}^{t} dt = t. \tag{4.5}$$

The "$y =$" notation emphasizes that the limits of integration are stated in terms of a variable different from the variable of integration. This is a useful trick that can prevent much grief whenever you use substitution in integration.

Integrating Eqn. 4.5 yields

$$[\log u]_{y=y_0}^{y=y} = bt$$

$$[\log(a + by)]_{y_0}^{y} = bt$$

$$\log(a + by) - \log(a + by_0) = bt$$

$$\log \frac{a + by}{a + by_0} = bt$$

$$\frac{a + by}{a + by_0} = e^{bt}$$

$$a + by = (a + by_0)e^{bt}$$

$$y = \frac{1}{b}\left[(a + by_0)e^{bt} - a\right], \text{ or } y = [(a/b) + y_0]e^{bt} - (a/b). \tag{4.6}$$

This is the solution of all differential equations of the form

$$y' = a + by; \quad y(0) = y_0, \tag{4.7}$$

as long as $b \neq 0$. Before we apply it to our three special cases though, it would be wise to check it. *In fact, whenever you obtain a solution to a differential equation, it is advisable to check it!*

4.3   Checking Solutions of ODEs

An analytical check is based on the fact that if y is the solution of the equation $dy/dt = f(y)$, then it must be true that

derivative of solution $y(t) = f(\text{solution } y(t))$.

With our present equation, that becomes

derivative of solution $= a + b \times$ solution.

The solution at the initial value of t must also equal the initial condition. The complete check thus involves two steps, namely:

- At $t = 0$, y should equal y_0. Does it?

- Since our solution is $y(t)$, we should be able to show that its derivative is equal to the original differential equation, for all y. We attempt that now.

For the moment, let us denote our putative solution as $s(t)$; i.e.,

$$s(t) = \frac{1}{b}[(a + by_0)e^{bt} - a]. \tag{4.8}$$

Differentiating this solution yields

$$s'(t) = \frac{d}{dt}\left\{\frac{1}{b}[(a + by_0)e^{bt} - a]\right\}$$

$$= \frac{d}{dt}\left[\frac{1}{b}(a + by_0)e^{bt}\right] - \frac{1}{b}\frac{da}{dt}$$

$$= \frac{a + by_0}{b}\frac{de^{bt}}{dt} - 0 = \frac{a + by_0}{b}be^{bt},$$

or $s' = (a + by_0)e^{bt}$.

Further, if $y = s(t)$ is a solution of our original DE, then it must also be true that

$$s' = a + bs.$$

If we substitute $s(t)$ in the RHS of this latter equation, we obtain

$$s' = a + b\left\{\frac{1}{b}[(a + by_0)e^{bt} - a]\right\}, \text{ or}$$

$$s' = a + (a + by_0)e^{bt} - a = (a + by_0)e^{bt}.$$

But this is identical to s' obtained by differentiating $s(t)$ (Eqn. 4.8). Thus, s *must* be a solution, as long as the initial condition is also satisfied[2].

To summarize how to check a solution in the general case:

- begin with a differential equation $y' = f(y,t)$ and find a supposed solution, $y = s(t)$.

- Substitute s for y in $y' = f(s,t)$.

- Differentiate the solution $s(t)$ to get $y' = s'(t)$.

- Determine whether $f(s,t) = s'(t)$—if so, the solution checks for the general case.

- Finally, $s(0)$ must equal the initial condition y_0.

The reader should be sure to understand the steps of this check, so as to be able to carry out similar checks for solutions of other differential equations.

An alternative method for checking a putative solution to a differential equation makes use of numerical differentiation, which we developed in Section 2.4, p. 32. This method may be easier, but is not a definitive proof of correctness. To demonstrate this method, let us again check whether Eqn. 4.8 is a solution to Eqn. 4.7. To work numerically, we must replace all symbols with numbers—I recommend using a different prime number for each quantity, to help prevent spurious cancelling of errors.

For example, let us replace $a = 2$ and $b = 3$ in Eqn. 4.7, to yield $y' = 2 + 3y$, and set $y_0 = 5$. Then we wish to show that

$$s(t) = \frac{1}{b}[(a + by_0)e^{bt} - a] = \left\{[17e^{3t} - 2]/3\right\} \qquad (4.9)$$

is a solution. If it *is*, then with t set to some arbitrary value ($t = 7$, say) and for some small h ($h = 0.001$, say) it should be true that

$$\frac{ds}{dt} \approx \frac{s(t+h) - s(t-h)}{2h} \approx a + bs(t),$$

where the middle expression is the central-difference approximation to the derivative, and the right-most expression is the RHS of the original differential equation with y replaced by the supposed solution, s.

---

[2]Theoretical considerations outside our scope show that a first-order linear DE like this has a single unique solution, so this is not only *a* solution but *the* solution.

If we now substitute the RHS of Eqn. 4.9 for s, with its three different arguments of $t + h$, $t - h$, and t, we obtain

$$\frac{\left\{[17e^{3(7.001)} - 2]/3\right\} - \left\{[17e^{3(6.999)} - 2]/3\right\}}{0.002} \quad ? = ?$$

$$2 + 3\left\{[17e^{3(7)} - 2]/3\right\}.$$

The calculations indicated yield $2.24199 \times 10^{10} \approx 2.241987 \times 10^{10}$, which is reasonable agreement, probably within round-off error.

This calculation provides some confidence that Eqn. 4.8 is a solution to Eqn. 4.7, but it does not serve as a proof—the analytic check above is distinctly more definitive. Repeating the calculations with other numerical values (and with various combinations of positive and negative numbers) would improve our confidence, however. I leave it to the reader to repeat this check with a different set of coefficients, and in particular, with a negative value for b.

The Mercury Problem

Finally, we are ready to apply our general solution to the more specific $m'(t)$, $D'(t)$, and $c'(x)$ equations. For the mercury in the lake, we had

$$m'(t) = q\left(c_i - \frac{m}{V}\right); \quad m(0) = m_0.$$

This is equivalent to $y' = a + by$ if $a = qc_i$ and $b = -q/V$. We substitute these values into the general solution,

$$y = \frac{1}{b}\left[(a + by_0)e^{bt} - a\right]$$

to obtain

$$m(t) = -\frac{V}{q}\left[\left(qc_i - \frac{q}{V}m_0\right)\exp(-qt/V) - qc_i\right].$$

A little algebra simplifies this to

$$m(t) = (m_0 - Vc_i)\exp\left(-\frac{q}{V}t\right) + Vc_i.$$

If we set $m_0 = 75$ kg, $V = 3 \times 10^6$ m^3, $c_i = 300$ mg m^{-3}, and $q = 912500$ m^3 yr^{-1} (corresponding to the problem posed on p. 12), the solution appears as in Fig. 4.4, p. 97. This solution is correct at $t = 0$ and seems reasonable as $t \to \infty$, when the lake concentration approaches that in the incoming stream.

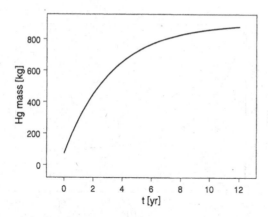

Figure 4.4: A solution curve for mercury entering a lake.

The Swallow's Flight (or Plight)

To see what happens to the swallow we left suspended in mid-air a few pages back, recall that we had $D' = H - D/4$, with $D(0) = D_0$. Our general solution will apply if we substitute $y = D$, $a = H$, and $b = -1/4$. Then $y = [(a/b) + y_0]\exp(bt) - a/b$ becomes

$$D(t) = (D_0 - 4H)e^{-t/4} + 4H.$$

Again, if $t = 0$, $D = D_0$ as it should. However, as time goes on, the exponential factor approaches zero, so $D(t)$ approaches $4H$ at large t. In other words, because of her unusual behavior, this particular swallow can never get closer to the mission than 4 km for each km da^{-1} of headwind speed. If the headwind is at all substantial, the tourists will never see her. In any case, our swallow's time-distance curve is as shown in Fig. 4.5, p. 98, if she starts 4000 km from the mission, and if the headwind is 250 km da^{-1}.

Solids in the Stream

I leave it to the reader to manipulate our general solution so that it yields $c(x)$ for the solids concentration problem, which was

$$\frac{dc}{dx} = -\frac{k}{u}c.$$

Be sure to sketch your solution, and to show that it gives reasonable results at $x = 0$ and at very large distances downstream from the source.

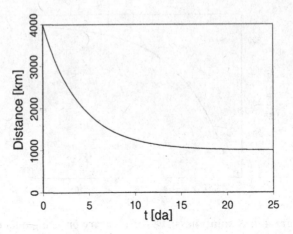

Figure 4.5: A solution curve for the swallow's distance from the mission.

4.4   Notes on Differential Equations

Having considered three examples of the *linear, first-order, ordinary differential equation with constant coefficients*, we now define differential equations more formally. We then consider more examples, and take up the analytic solution of DEs that can be solved by separation of variables. Later, in Chapter 7 we will consider methods for *numerical* solution of ordinary differential equations that can't be solved analytically.

What Are Differential Equations?

Definition: An *ordinary differential equation* (ODE) is any equation involving one or more ordinary (i.e., not partial) derivatives. Thus, an nth-order ODE is any function of the form

$$F\left(\frac{d^n y}{dx^n}, \frac{d^{n-1} y}{dx^{n-1}}, \ldots, \frac{dy}{dx}, y, x\right) = 0,$$

and a solution of this equation is any function $y = f(x)$ for which $F = 0$. This equation is nth-order because that is the highest order derivative appearing in it. The independent (x) variable is often time (t) in practice.

In contrast to ODEs, there are *partial differential equations* (PDEs) that involve partial derivatives. One example, which describes time-dependent heat transfer in a three-dimensional medium such as a volume of soil, is

$$\frac{\partial^2 T}{\partial x^2} + \frac{\partial^2 T}{\partial y^2} + \frac{\partial^2 T}{\partial z^2} = \alpha \frac{\partial T}{\partial t}.$$

Here, T is soil temperature; x, y, and z are coordinates in the 3-dimensional volume; t is time; and α is a coefficient that accounts for the thermal properties of the soil. Partial derivatives are required because temperature varies with both time and location in the soil. We will take up PDEs in Chapter 11.

The *order* of an ODE is the order of the highest derivative in it. Thus the logistic equation for population growth,

$$\frac{dN}{dt} = rN\left(\frac{K - N}{K}\right) = rN - \frac{r}{k}N^2,$$

is first-order (although non-linear because of the term in N^2), while

$$a\frac{d^2 y}{dt^2} + b\frac{dy}{dt} + cy = f(t)$$

is an important second-order equation from physics that describes many phenomena including both heat and mass transfer, and periodic responses like swinging pendula and alternating current. We will concentrate on first-order ODEs for the rest of this chapter, and take up second-order ones in Chapter 8.

Many important DEs are of the form $y' = f(y)$. That is, the state variable y changes at a rate that depends only on the current value of the state. Examples are the logistic equation and our form $y' = a + by$. Other DEs can be of the form $y' = f(t, y)$, with the rate depending on both the driving and response variables. An example of the latter would be a modification of the logistic equation in which the carrying capacity was a periodic (*e.g.* seasonal) function of time; i.e., $K = a + b \sin ct$:

$$\frac{dN}{dt} = rN\left[\frac{(a + b \sin ct) - N}{(a + b \sin ct)}\right] = f(N, t).$$

The form $y' = f(t)$, such as $y' = kt$, can exist, but seldom arises in practice. (The landfill PCB example from p. 53 was an example of such an equation, however.)

Another classification important in the theory of DEs, is the distinction between linear and non-linear equations. A first-order ODE is linear if y and y' each appear (if at all) only to the first power and not multiplying each other. Thus $f_1(x)y' = f_2(x)y + f_3(x)$ is the most general linear, first-order ODE. This equation is considered linear in y even if the functions f are non-linear in x. (*Solutions* to such equations are often non-linear, as you have seen.) Note that the simple form of the logistic equation is non-linear; this may be easier to see if you write it as $N' = rN - rN^2/K$. The term in N^2 makes this equation non-linear.

A Graphical View of ODEs and Their Solutions

Graphical interpretations of mathematical concepts often provide useful aids to understanding, at least for some people. Treating an integral as an area under a curve, or a derivative as the slope of a curve, are good examples. Here we take a graphical look at differential equations.

Consider first the simple DE,

$$\frac{dy}{dx} = x,$$

with no initial condition specified. This equation is a rule stating that if you were at any point $[x, y]$ in the x-y plane, you would have to move away from that point along a line with slope x if you wanted to satisfy the equation.

Of course, you could move along that line for no more than an infinitesimal distance—as soon as you moved a yoctometer (10^{-24} m) or so, the value of x would have changed a little bit, and you would have to change your direction of movement a little bit too.

Fig. 4.6 (left), p. 101, which represents $y' = x$, can be interpreted in several ways. First note that for any of the curves suggested by the figure, the slope of the curve at any value of x is equal to x. For example, at $x = 1$, each of the potential curves has a slope (dy/dx) equal to 1, regardless of the value of y. Above $x = -2$, each curve has a slope of -2, and above $x = 0$, each curve has a slope of zero. (Note that the curves actually plotted in the figure are only representatives of all the infinity of curves that could be plotted. You might imagine that each pair of curves shown has a thousand, or a million, or a billion others drawn between them in invisible ink.) For another interpretation, imagine that you parachuted onto some point on a huge

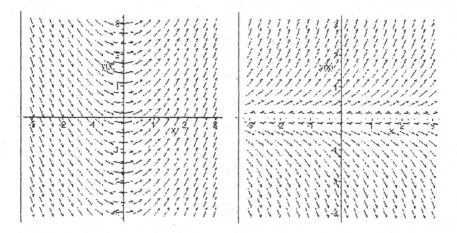

Figure 4.6: Direction fields for the differential equation $y' = x$ (left), and for $y' = y$ (right).

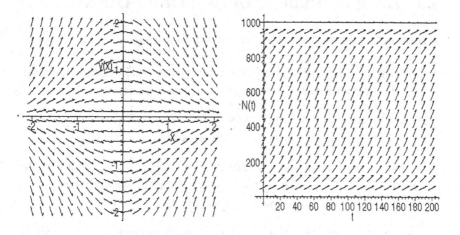

Figure 4.7: Direction fields for the differential equation $y' = -xy$ (left), and for the "logistic equation" with $r = 0.05$ and $K = 1000$ (right).

copy of this figure (with x and y having units of kilometers, say). Your landing point would be your *initial condition* (or IC). Given that IC, you would have to walk along whatever curve (visible or invisible) that you had landed on to satisfy the DE, $y' = x$. In other words, the curves represent various possible *solutions* of the DE, depending on what the IC might be. Once the IC is specified, however, only *one* curve represents a valid solution.

More examples might help. Figs. 4.6 (right) and 4.7 (left), p. 101, represent the *direction fields* (solution fields) of the DEs $y' = y$ and $y' = -xy$, respectively. In the former case, the slopes of the solution curves depend only on y, and for fixed y they do not vary with x. In the second of these figures, the slopes depend on both x and y. For example, if either $x = 0$ or $y = 0$, the slope is zero. And, the slope would be +1 for (x, y) equal to (-2, +0.5), (-1, +1), (-0.5, +2), (+0.5, -2), (+1, -1), (+2, 0.5), or any other similar combination. Check this out by sketching in the curves through those points.

As a final example, Fig. 4.7 (right), p. 101, traces representative solution curves for the logistic equation,

$$\frac{dN}{dt} = rN\left(\frac{K - N}{K}\right).$$

4.5 Analytic Solution of First-Order ODEs

Several points about solving DEs are worth noting:

- Solving a DE is often called "integrating the DE," because integrating dy/dx with respect to x gives y, the solution sought.

- Like integration, solving DEs is not always straightforward. Analytic solution of DEs is not always possible.

- If you *do* obtain an analytic solution of the form $y = F(x)$ to the DE $y' = f(x, y)$ with $y(0) = y_0$, then it is straightforward to check your solution using the steps outlined earlier.

- As we shall see, DEs that can't be solved analytically can still be solved numerically. However, numerical solutions have three major disadvantages:

  1. Each such solution applies to only one special case, with a particular initial value and particular numerical values for the parameters (constants).
  2. Numerical solutions are difficult or impossible to check rigorously.
  3. Numerical solutions can have large computational errors, due both to truncation and to round-off. We'll take up numerical methods soon, for use when analytic solutions are not available.

Solution by Separation of Variables

Although there are many methods for solving first-order ODEs analytically (especially for linear equations), most are beyond the scope of this book. For now, we will take up only one straightforward method known as separation of variables, a method that can be used for a special class of DEs of the form

$$F(x, y) = \frac{dy}{dx} = f(x) \times g(y), \text{ with } y(x_0) = y_0. \tag{4.10}$$

(Separation does *not* work for equations like $dy/dx = f(x) + g(y)$.) For equations of the product form shown in Eqn. 4.10, one can multiply both sides by dx and divide both sides by $g(y)$ to obtain

$$\frac{dy}{g(y)} = f(x)dx,$$

as we did with Eqn. 4.4, p. 92.

We have now accomplished the separation, with all y dependence on one side and all x dependence on the other. Next we attempt to integrate both sides:

$$\int \frac{dy}{g(y)} = \int f(x)dx.$$

If both these integrals can be evaluated analytically, we will have our solution. A new example may help:

In economics, it has long been noted that a dollar is worth more to the poor (e.g., college professors) than to the wealthy (college administrators). A simple model of this notion states that one's rate of increase in *utility* (U) for a unit increase in *wealth* (W) is inversely proportional to one's present wealth. Thus

$$\frac{dU}{dW} = \frac{k}{W}, \text{ with } U(W_0) = U_0.$$

Here k is the constant of proportionality for a particular person. Separating and integrating this model with the IC specified in the limits of integration yields

$$\int_{U_0}^{U} dU = k \int_{W_0}^{W} \frac{dW}{W}$$

$$\left[U \right]_{U_0}^{U} = k \left[\log W \right]_{W_0}^{W}$$

$$U - U_0 = k(\log W - \log W_0)$$

$$U = U_0 + k \log\left(\frac{W}{W_0} \right).$$

The IC can be incorporated into the solution process in an alternate way that uses indefinite integration and constants of integration, instead of including it in the limits of integration. Again we separate variables and integrate:

$$\int dU = k \int \frac{dW}{W}$$

Both integrals produce a constant of integration, but these can be combined into one constant C. Thus, integration yields a general solution,

$$U = k \log W + C. \tag{4.11}$$

The solution must hold at the IC as well as elsewhere, so $U_0 = k \log W_0 + C$, from which

$$C = U_0 - k \log W_0.$$

Substituting this expression for C in the general solution (Eqn. 4.11) yields

$$U = k \log W + U_0 - k \log W_0,$$

which, with a little algebra, is identical to the solution 4.11 obtained by the first method. Using the IC in the limits of integration (rather than to solve for a constant of integration) is generally simpler, but you may want to try both methods to see which works best for you.

Solution Using Tables

Although separation of variables is an important solution method for many first-order DEs, it only applies to those equations that can be factored into the form $y' = f(x)g(y)$. Thus, it can be used for $y' = xy$ or $y' = e^x \sin y$, but not for $y' = x + y$ nor for $y' = e^{-xy}$. The first of these can be solved using an integrating factor (Chapter 5), but some DEs require even more complex methods, or can't be solved analytically at all. Because other analytic solution methods (e.g., Derrick and Grossman 1981) are beyond the scope of this book, we next take up how to work with tables of ODE solutions.

A comprehensive book of tables, e.g., Murphy (1960), can be useful if you have a recalcitrant ODE to solve. Unfortunately, that book is out of print (although it is available in many libraries). Section 4.6, p. 106, provides a short table of some of the DE forms that arise most often in practice. The table contains eleven differential equations of the general form $y' = f(x)$. Using it to solve a particular equation requires

- Identifying the dependent (y) and independent (x) variables in your own equation.

- Finding a comparable form in the table, with all the needed terms. For example, if your equation contains a linear term in y (e.g., $5y$) and an exponential term in x (e.g., $2e^{-3x}$), then you look for an equation in the table that contains terms of those same forms.

- Substituting your variables for the x and y in the table, and your parameter values (constants) for those in the table.

Some of the solutions provided require additional steps, as we will illustrate.

Suppose you have the equation $N' = 3 - e^{-t/2} + 0.02N$, with the IC $N(1) = 50$. Our "x" variable is thus t and our "y" variable is N. This DE includes a constant term, an exponential in "x," and a linear term in "y." A search through the table yields equation number 10 of the form $y' = a + be^{kx} + cy$, which applies when $k \neq c$. We set up a list to match variables and constants; i.e., $y \rightarrow N$, $x \rightarrow t$, $a \rightarrow 3$, $b \rightarrow -1$, $k \rightarrow -0.5$, and $c \rightarrow 0.02$. Note that k and c *are* different, so we can use equation number 10.

The table lists two solutions. The first, Eqn. 4.12, applies when the IC is defined at $t = 0$, and the second, Eqn. 4.13, applies for a more general IC of form $y(x_0) = y_0$. Because our present IC is $N(1) = 50$, we must use this second form. Thus, our solution must be

$$y(x) = \frac{a(c-k) + [a(k-c) + cky_0 - c^2y_0]e^{c(x-x_0)}}{c(k-c)} +$$

$$\frac{-bce^{cx-cx_0+kx_0} + bce^{kx}}{c(k-c)}.$$

Substitution of values from our list then yields

$$N(t) = \frac{3(0.02+0.5)}{0.02(-0.5-0.02)} +$$

$$\frac{[3(-0.5-0.02) - 0.5(0.02)50 - 0.02^2 50]e^{0.02(t-1)}}{0.02(-0.5-0.02)} +$$

$$\frac{0.02e^{0.02t-0.02-0.5} - 0.02e^{-0.5t}}{0.02(-0.5-0.02)}.$$

Finally, a bit of arithmetic simplifies this to

$$N(t) = \alpha + \beta e^{0.02(t-1)} - \gamma e^{0.02t-0.52} + \gamma e^{-0.5t},$$

where $\alpha = -150$, $\beta = 200$, and $\gamma \approx 1.923076923$. The computations required to arrive at this point have been sufficiently messy that we would surely want to check this solution by one of the methods described in §4.3, p. 94. That exercise is left to the reader.

Although separation of variables, other analytic methods, and tables like Murphy's will allow you to solve many ODEs, there remain many others that simply have no analytic solution. This is especially true for *systems* of ODEs, which we take up in Chapter 6. For this reason, we will need to take up numerical methods for solving ODEs or systems of ODEs, and we will do this in Chapter 7.

4.6 Table of Solutions of Selected ODEs

This table contains solutions to some common first order, ordinary differential equations $[y' = f(y,t)]$ that can't be readily solved by separation or similar means. Solutions containing "y_0" assume the initial condition $y(0) = y_0$. Solutions containing a constant of integration C apply when the initial condition is specified at a non-zero time—for such cases, the value of C must be calculated from the initial condition. Both forms are given for some, but not all, of the equations treated here.

The solutions here were obtained using Maple V, Release 4, student edition. They are presented in a form similar to that in Murphy (1960). That excellent set of tables is unfortunately out of print, but is still available in some science-oriented libraries. I have not found any similarly comprehensive book that is still in print.

1. $y' = a + by$:

$$y(t) = \left(\frac{a}{b} + y_0\right) e^{bt} - \frac{a}{b}; \quad y(t) = C e^{bt} - \frac{a}{b}$$

2. $y' = a + bt + cy$:

$$y(t) = \left(y_0 + \frac{a}{c} + \frac{b}{c^2}\right) e^{ct} - \left(\frac{a}{c} + \frac{b}{c^2} + \frac{bt}{c}\right)$$

$$y(t) = C e^{ct} - \frac{ac + b + bct}{c^2}$$

3. $y' = a + by/(c + t)$:

$$y(t) = \frac{(c + t)^b}{c^b}\left[\frac{ac}{(b-1)} + y_0\right] - a\frac{(c + t)}{(b - 1)}$$

$$y(t) = C (c + t)^b - \frac{a(c + t)}{(b - 1)}$$

4. $y' = ay^2 + by + c$. Solving this equation takes a little extra effort. You first have to set the right-hand side (the quadratic) to zero and find its roots using the quadratic formula. Call these roots y_1 and y_2. Then, if $y_1 = y_2$, rewrite the DE as $y' = a(y - y_1)^2$. The solutions for this form are:

$$y(t) = \frac{y_1\left[at(y_1 - y_0) + 1\right] - (y_1 - y_0)}{at(y_1 - y_0) + 1}$$

$$y(t) = y_1 - \frac{1}{at - C}.$$

If the two roots are distinct (i.e., $y_1 \neq y_2$), then rewrite the DE as $y' = a(y - y_1)(y - y_2)$. The solutions for this form are

$$y(t) = \frac{y_2 - y_1 \left(\dfrac{y_2 - y_0}{y_1 - y_0}\right) \exp\left[(y_2 - y_1)at\right]}{1 - \left(\dfrac{y_2 - y_0}{y_1 - y_0}\right) \exp\left[(y_2 - y_1)at\right]}$$

when the IC is given at $t = 0$, or

$$y(t) = \frac{y_2 - y_1 \exp\left[(y_2 - y_1)(at - C)\right]}{1 - \exp\left[(y_2 - y_1)(at - C)\right]}$$

otherwise.

5. $y' = ax^n y$:

$$y(t) = y_0 \exp\left(\frac{at^{n+1}}{n+1}\right)$$

$$y(t) = C \exp\left(\frac{at^{n+1}}{n+1}\right).$$

6. $y' = a^2 + b^2 y^2$. (Expressing the constants as squares simplifies the form of the solution.)

$$y(t) = \frac{a}{b} \tan\left[abt + \arctan\left(\frac{b}{a}y_0\right)\right]$$

$$y(t) = \frac{a}{b} \tan[ab(t - C)]$$

7. $y' = a^2 - b^2 y^2$:

$$y(t) = \frac{a\left[(a + by_0) e^{2abt} - (a - by_0)\right]}{b\left[(a + by_0) e^{2abt} + (a - by_0)\right]}$$

$$y(t) = \frac{-a}{b} \cdot \frac{1 - \exp\left[2ab(t - C)\right]}{1 + \exp\left[2ab(t - C)\right]}.$$

8. $y' = (t + y)^2$:

$$y(t) = \frac{t\cos t - \sin t - y_0(t \sin t + \cos t)}{y_0 \sin t - \cos t}.$$

9. $y' = a\cos(bt + c) + ky$:

$$y(t) = y_0 e^{kt} + \frac{a}{b^2 + k^2} \times$$

$$\left\{ [b \sin(bt + c) - k \cos(bt + c)] - [b \sin c - k \cos c] e^{kt} \right\}.$$

10. $y' = a + be^{kx} + cy, k \neq c$:

If $y(0) = y_0$, then a solution (not necessarily in simplest form) is

$$y(x) = \frac{\{a(k - c) + c[(k - c)y_0 - b]\} \exp(cx)}{c(k - c)} \tag{4.12}$$

$$+ \frac{bc \exp(kx) - a(k - c)}{c(k - c)}.$$

If $y(x_0) = y_0$, then more generally,

$$y(x) = \frac{a(c - k) + [a(k - c) + cky_0 - c^2 y_0] e^{c(x - x_0)}}{c(k - c)} \tag{4.13}$$

$$+ \frac{-bce^{cx - cx_0 + kx_0} + bce^{kx}}{c(k - c)}.$$

11. $y' = a + be^{cx} + cy$:

If $y(0) = y_0$, then

$$y(x) = \frac{(bxc + a + y_0 c) e^{cx}}{c} - \frac{a}{c}.$$

If $y(x_0) = y_0$, then more generally,

$$y(x) = b(x - x_0) \exp(cx) + \frac{(a + cy_0) \exp[c(x - x_0)] - a}{c}.$$

4.7 Analytic Solution with MATLAB

It is often convenient to use software like MATLAB®to search for analytic solutions to ODEs. As an example, consider again our familiar logistic population-growth equation,

$$\frac{dN}{dt} = rN\left(\frac{K - N}{K}\right) = rN - \frac{r}{k}N^2.$$

The solution, which was given in the introduction (p. 9), could be obtained from the fourth equation in the table of §4.6. However, that process is fairly complex, and if MATLAB or a similar program is available, it may be easier to obtain the solution that way. Here's how, assuming MATLAB's Symbolic Math Toolbox is available:

The command

```
dsolve('DN=r*N*(K-N)/K')
```

supplies a solution including an undefined constant of integration, because no IC is given. To supply one, enter something like

```
N=dsolve('DN=r*N*(K-N)/K','N(0)=N0').
```

That yields the more convenient solution,

```
N=K/(1+exp(-r*t)*(K-N0)/N0).
```

What happens if we ask MATLAB to solve an ODE that has no analytic solution (or one it is not capable of solving, at any rate)? If we enter

```
dsolve('Dy=sin(exp(-y^2/2))'),
```

MATLAB returns

```
ans=t-Int(1/sin(exp(-1/2*_a^2)),_a = .. y)+C1 = 0.
```

The reference to "Int" indicates that the solution involves an integral that MATLAB can't perform analytically. (Note that this is not a very realistic equation, but it *is* one MATLAB can't solve.)

4.8 Exercises

In these exercises, be sure as always to check units frequently, and when you solve a differential equation, check your solution as well.

1. Show that $de^t/dt = e^t$ (the function e^t is its own derivative) by differentiating the Maclaurin series for e^t (p. 28) term by term, and showing that the result is the same as the original series.

2. Suppose that during a three-month drought, the streamflow q_i into a western lake declines over time, specifically as $q_i = q_{i0} \exp(-kt)$, where q is in m^3 da^{-1}, and t is in da after the beginning of the drought.
   (A) What must be the units of k?

Also, suppose the shape of the lake basin and its outlet are such that the outflow q_o [m^3 da^{-1}] depends on the volume V [m^3] of water present in the lake according to $q_o = a + bV + cV^2$. Further, suppose that about E m^3 of water evaporates from the lake's surface each day.

(B) What must be the units of a, b, and c?

(C) Derive the equation that, if solved, would allow you to calculate the volume of water in the lake at any time during the period in which these conditions hold.

(D) Can your equation be solved for $V(t)$ by separation? If so, do so. If not, check whether MATLAB can find an analytic solution.

3. Two streams, A and B, enter a lake containing water volume V m^3, from which a third outlet stream, C, leaves. Stream A has a flow rate (assumed constant) of q_a m^3 da^{-1}, and it carries a concentration c_A mg m^{-3} of arsenic (As) from mine tailings that it flows past. Stream B has a constant flow rate of q_B m^3 da^{-1}, and it carries a negligible amount of arsenic. The flow in the outlet C is just the sum of the flows of the inlets, causing the volume of water in the lake to remain constant. As an approximation, assume the water in the lake is uniformly mixed at all times. A fraction f of the As in the lake water falls to the bottom of the lake each day, as a result of uptake by organisms that then die and fall to the sediment.

   A. Derive the ODE describing how the mass of As in the lake water changes with time, as a result of those processes.

   B. Using any convenient method, provide the solution to that equation. Assume that the initial As content of the lake is m_0 mg.

4. A lump of a radioactive material loses mass at an instantaneous rate that is a fraction f per year of its mass at any given time. If the lump has an initial mass m_0 g at time 0, what is the equation for its mass at later times? Obtain that by first writing and then solving an appropriate differential equation.

5. A consulting company, EcoTox, is testing a new insecticide for its manufacturer. One test they perform involves supplying a pulse of the insecticide to a large fish tank. Specifically, the tank contains a constant volume V [m^3] of water, and water flows into and out of

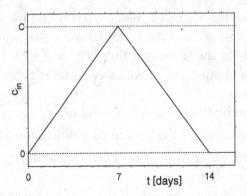

Figure 4.8: Illustration of the triangular pulse of insecticide concentration in the water entering the tank.

the tank at a constant rate q [m^3 da^{-1}]. They vary the concentration of the insecticide flowing into the tank in a triangular pulse of two-week duration, as shown in Fig. 4.8. As shown there, the ingoing insecticide concentration c_{in} [g m^{-3}] starts at zero, rises to C_{max} (shown as C in the figure) at the end of the seventh day, and drops back to zero at the end of the fourteenth day and thereafter.

Your task is to determine how the insecticide concentration c_t *in the tank* changes over those same 14 days, neglecting any potential uptake by the biota or other potential losses. That is, consider only the flows into and out from the tank, and what stays in the water there. Also assume that the contents of the tank are always well mixed.

To accomplish that task, derive and solve one differential equation that applies for the first seven days, assuming that $c_t = 0$ at the beginning of the entire process. Then, derive and solve a slightly different DE that applies over days 8-14. Note that the solution of the first equation at $t = 7$ days is the initial condition for the concentration in the tank when the second (declining) stage of the pulse begins.

You may leave an undetermined constant in the second solution, but not in the first. Be sure to specify the initial condition for the second DE as specifically as you can.

6. Suppose you are working with a gas analyzer that measures SO$_2$ concentrations in air samples. The instrument contains a stainless

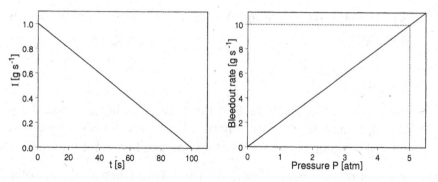

Figure 4.9: Inflow rate I [g s^{-1}] versus time t [s] after a "sample" button is pushed (left), and Bleedout rate B versus pressure P (right).

steel flask of 1 liter capacity. Gas is pumped into one end of the flask at an input rate I g s^{-1}, which is controlled electronically to be

$$I = 1 - t/100 = 1 + at \quad \text{for} \quad 0 \le t \le 100 \text{ s},$$

$$I = 0 \quad \text{thereafter.}$$

This varying inflow rate is shown in Fig. 4.9 (left).

Gas bleeds out of the flask (into a measuring chamber) at a rate B [g s^{-1}] that is proportional to the pressure in the flask [atm], as defined graphically in Fig. 4.9 (right).

The temperature in the flask is held constant, and as a result the pressure P in the flask is proportional to the mass M of the gas it contains. Specifically, for this particular gas, we know that $P = cM$ with $c = 0.83$ atm g^{-1}.

A. If the flask initially contains 1.7 g of the gas, derive the differential equation relating M to t for $0 \le t \le 100$.

B. Can this equation be solved by separation? If so, do so. If not, explain why and solve it by another method.

7. Solve the following equations by separation of variables. Each has the IC $y(x_0) = y_0$: $y' = -3y$, $y' = 2x^2$, $y' = -7xy$, and $y' = \sin 3x$.

8. Use the methods of p. 95 to check whether $y = b \exp(1.5t^2)$ is the solution to the differential equation $dy/dt = 3ty$ if $y(0) = b$. Don't forget to check whether the initial condition is satisfied.

9. Use the tables to find the solution to the logistic population growth equation,

$$\frac{dN}{dt} = rN\left(\frac{K-N}{K}\right),$$

   when $N(0) = N_0$. Hint: To find the right form in the table of solutions, it may help to multiply through by rN. This solution was given in Chapter 1, but see whether you can find it for yourself.

10. Consider $dy/dx = 1/(e^{2y}x^2)$ with $y = 1/10$ when $x = 1/2$. Solve this equation analytically for $y = f(x)$, and calculate the value of y when $x = 0.9$. The solution should be in a form such that if given an x value you could provide the corresponding y value.

11. A basement crawl space has a volume V m^3 and a floor area A m^2. Radon gas infiltrates through the dirt floor at a rate r μg m^{-2} da^{-1}. Air from outside, essentially radon-free, enters the space through a ventilator at a rate q m^3 da^{-1}, and the same volume of air (carrying the present concentration of radon with it) flows out the other side. Finally, some of the radon present is lost by radioactive decay at an instantaneous fractional rate of f da^{-1}. Derive and solve the differential equation describing this process. You may assume as an initial condition that no radon is present.

12. Lake B in the chain diagrammed in Fig. 4.10 receives water via the stream from A, and discharges water into the stream to C. Discharge from A (flow into B) is regulated by a hydroelectric dam and has the form $[a + b\sin(t/2\pi)]$ where the units of a and b are m^3 day^{-1}, and t is in days.

   The outlet of lake B is essentially wedge shaped (Fig. 4.11, p. 115).

   Water flows from this outlet at a rate proportional to the area available for flow (i.e., to the shaded area shown). The basin of

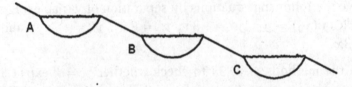

Figure 4.10: Three lakes (A, B and C) connected by a stream.

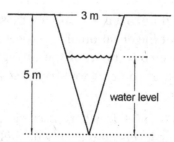

Figure 4.11: Cross-section of outlet of lake B.

B is hemispherical, such that the relation between volume (V) of water in the basin and water level (W) is

$$V = \pi \left(5W^2 - \frac{W^3}{3} \right),$$

with V in m^3 and W in m. (Be glad you weren't asked to derive that; try it for week-end entertainment!) Also, the surface area of Lake B is $\pi(10W - W^2)$.

If the water level of Lake B is W_0 at some reference time $t_0 = 0$, derive an equation which, if solved, would give you $W(t)$ in the future. You don't have to solve the equation. Assume a symbol for each constant you may need. State the units of every quantity. It will help to remember that the area of a right triangle is $1/2$ the base times the height; the wedge-shaped outlet shown the figure can be treated as two right triangles.

13. On p. 103, I claimed that $U = U_0 + k \times \log(W/W_0)$ was the solution to the wealth-utility equation, $dU/dW = k/W$. Show that this checks. Don't forget to check the IC, too.

14. A team of aquatic ecologists interested in nitrogen dynamics build a series of mesocosms (large aquaria) in which they set up artificial ecosystems including phytoplankton, zooplankton, and a small starting population of fathead minnows. They provide the mesocosms with limited nitrogen, but with plenty of all other required nutrients.

The ecologists decide to test the following model for the mass of nitrogen $M(t)$ (in g) that is stored in the biomass of the minnows, as the minnow population feeds and increases through time:

- The total mass of nitrogen N in a given mesocosm remains constant throughout the duration of an experiment. Any processes that might cause conversion of the nitrogen in the system to gaseous forms that could be lost from the system are assumed to be negligible.

- At all times, $N = M + L$, where L is the mass of nitrogen in the lower trophic levels (and everywhere else in the mesocosm other than in the fish biomass).

- Minnow biomass is represented for the purposes of the model by the mass of nitrogen in the fish (i.e., by M), rather than by the total mass of the fish.

- The minnows take up nitrogen at a rate that is proportional to the product of their current biomass (in nitrogen units) and of the amount of nitrogen in the rest of the ecosystem.

- The initial nitrogen mass in the minnows is M_0 g. Because N is constant, this leaves $L_0 = N - M_0$ as the initial nitrogen mass in the rest of the system.

Your job is to write the equation that, if solved, would give $M(t)$ at later times (assuming the model is correct). Provide clear definitions for all variables and constants that you use, and give units for each. Your final result should be entirely in terms of M, N, t, and any constants you may introduce—it should not contain the variable L.

15. Direction-field exercise

    The four direction fields in Figs. 4.12 and 4.13, p. 117, represent the four differential equations listed below. (The plots were produced with Maple's "dfieldplot" command.) Try to match each DE with one of the direction fields. In each plot, x ranges from 0 to 4, and y ranges from -2 to +2. Hint—consider where the slope should equal zero and where it should equal $\pm$ unity for each of these differential equations:

    i. $\dfrac{dy}{dx} = x + y$;   ii. $\dfrac{dy}{dx} = x - y$;   iii. $\dfrac{dy}{dx} = \dfrac{x}{y}$;   iv. $\dfrac{dy}{dx} = xy$.

16. Unlike our unfortunate swallow, the horses at Alec's Stable move faster as they approach home. Suppose at a distance $x = 2$ km

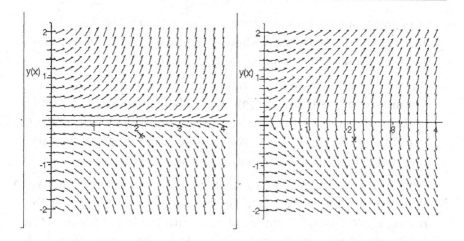

Figure 4.12: A (left), and B (right).

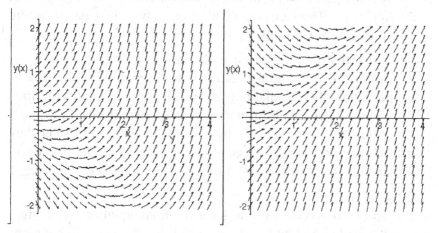

Figure 4.13: C (left) and D (right).

from home the horse you've rented is moving at a speed of s_2 km min^{-1} and at $x = 1$ km its speed is s_1 km min^{-1}. Also, suppose the speed increases linearly from $x = 2$ to $x = 0$. In terms of s_2 and s_1, how long will it take you and your mount to get to the stable, starting from the 2 km distance?

17. A couple is designing an energy-efficient home. It will include a heating system that, like most, is either off or on at any given instant. When it is on, it will supply Q J min^{-1} of heat. Whether the furnace is off or on is to be specified by a function $f(T)$ that will be determined by the thermostat they purchase, and is unknown

at present; however, the unitless function $f(T)$ will have values of either 0 or 1 depending on whether the furnace is off or on. Thus, the heat supplied to the house at any instant will be $Q f(T)$.

The house will lose heat at a rate [J min^{-1}] that is proportional to the difference between the internal room temperature (assumed uniform throughout the house) and the outside air temperature, both in deg C. This heat loss will also be inversely proportional to the thickness S [cm] of the insulation they will install.

Temperatures are related to heat flows in much the same way that concentrations are related to mass flows. Specifically for this house, a net flow of C Joules of heat into the living space of the house would raise its temperature by 1° C. This means that if dH (say) is a small change in the heat content of the house, then $dH/dT = C$. The constant C, with units here of J deg^{-1}, is the *heat capacity* of the house.

Assuming that the outside air temperature varies as $a + b \sin(rt)$, determine the differential equation that the couple should solve to estimate how the internal house *temperature* will vary with time. They plan to use this information to help them decide how much insulation to install, and whether the $f(T)$ of the controller they might buy is acceptable. Because $f(T)$ is not yet specified, you will not be able to solve this equation, so there is no need to attempt that.

Perform a complete unit check on the final equation. Your final result should be stated as an equation for the rate of change of temperature with time.

18. A late spring snowfall at 8000 ft elevation in the Colorado Rockies· falls steadily for just over two days at a rate of s g per m^2 of surface area per hour. Because it is fairly warm at ground level, whatever snow is present consolidates, melts, and runs off at an instantaneous rate of $M\%$ per hr.

   a) If the ground was free of snow at the start of this storm, what mass of snow was present per square meter at each time t while the above conditions held? Answer symbolically at this stage.

   b) Check that your solution obtained in Part (a) is correct.

c) If $s = 6$ g m^{-2} hr^{-1}, and $M = 2\%$ per hr, what mass of snow covers each square meter of surface 30 hours after the storm began?

4.9 Questions and Answers

1. Why is $y' = a + by$ called a linear differential equation when a graph of its solution isn't linear at all?

   • Linearity of a differential equation is defined by the nature of the derivative ($a + by$ in that one), and not by the nature of the solution. In $y' = a + by$, y' is a linear function of y, and so by definition, this *is* a linear differential equation. I agree with you, however, that the solution $[y(t) = \text{stuff} + \text{stuff} \times \exp(bt)]$ is definitely not a linear function of t. This is admittedly a little confusing.

   Even a differential equation like $x^3 y' = \sqrt{x} y + \cos(x)$ is linear in y, although it is far from linear in x. Because the right-hand side has y to no other power than the first, this is called a linear differential equation. I'm sure its solution, which would be y as some function of x, is not a linear function of x.

2. Please go over the steps for checking the solution to a differential equation again.

   • Here's an easy example that may help you to see the process. Suppose you have the differential equation $y' = y$, with an initial condition of $y(0) = A$. Now suppose I claim to you that the solution is $y(t) = Ae^t$. Here's the check:

   Step 1: At $t = 0$, is $y = A$? Answer: yes.

   Step 2: Is $y' = y$? Answer: $y' = Ae^t$ (from differentiating the $y(t)$ that I claim is the solution). And the right hand side of the DE, namely y, is also Ae^t. Therefore, if $y = Ae^t$, $y' = y$. So this checks too.

3. When using definite integrals to solve DEs, do you always integrate from 0 to t? Put another way, how do you choose the limits of integration?

• For any differential equation of the form $dy/dt = f(t) \times g(y)$, with initial condition $y(t_0) = y_0$, you separate variables, and then integrate the t side from t_0 up to general t. (Most commonly, t_0 is set to zero since starting time is usually arbitrary anyway. However, if in some problem you wanted to start at $t_0 = 1995$, say, then you would integrate from 1995 up to a general value of t.) Next, integrate the y side from y_0 up to the value of y that corresponds with the general t. Occasionally, the y_0 value is zero too, but not as often as the t_0 value is.

The wealth-utility equation was an example of this—it didn't start at zero wealth, but at some finite amount of wealth. So the lower limit of integration wasn't zero. (That's a good thing, since the log of zero doesn't exist.)

4. I'm trying to understand the application of ODEs in 'real life.' Does the swallow example suggest what really happens to birds?

• I guess something similar to that could happen (with head winds faster than the bird's speed), but doubt that it does very often. I used that example to illustrate instaneous rates and some common errors. It is mathematically similar to some real situations, but as stated is pretty fanciful. I've been in canoes when the head wind was faster than my paddling, although my paddling speed was not proportional to how far I was from my goal. Most of our other examples are very true to life, however.

5. Please explain again how linear DEs can result in non-linear graphs.

• Ok. With a linear differential equation, the *derivative*, dy/dx, (i.e., the *rate* at which y changes with x) is a linear function of *the response variable y*. The 'non-linear graphs' refer to the *solution*, $y = f(x)$. These are graphs of how y varies with *the independent variable x* (or time). These are entirely different types of relationships, so it's easy for one to be linear and the other not. Of course the two kinds of relationships are related, but one is linear and the other is not.

6. In working with the direction fields, we figured out how the graphs and equations were related by plugging in points. Is that the correct (and easiest) way to do it?

• I assume you mean to do something like "calculate the slope at

x = some value A, and at y = some value B", and then to see if the graph indicates that slope at that value of x and that value of y. If so, I don't know of a better way. Sometimes you can work with just one of x or y, as with $dy/dx = x/y$, which would produce an infinite slope wherever $y = 0$ (except perhaps where $x = 0$).

7. Was the purpose of the graphing exercise simply to show that slope is the derivative of the function, or are there other purposes?

• I see three purposes for working with direction fields. (1) To show that a differential equation without an initial condition represents a whole series of solutions, but that once you choose an IC, the solution is uniquely determined. (That's true for first-order linear equations, at least.) (2) To get you to think in terms of slopes and rates of change. (3) To give you more practice reading graphs (which is not easy for everyone), by thinking about them in a new way.

8. Are there other ways than numerical substitution to check solutions to DEs?

• Yes. Please reread the section on checking solutions. The first method, substituting the supposed solution $[s(t)]$ into both sides of the original DE and checking for equality, is analytic, and doesn't involve any numerical substitution at all. That method provides the only definite proof that you have a correct solution. A second method is to solve the DE by a numerical method (which we will take up later), and to check whether that numerical solution agrees approximately at various times with the analytic solution at those times.

9. (Comment) I'm understanding how to solve DEs pretty well, but could use more help setting them up.

• The "story problem" aspect of applied math, which is one of the main factors that separates it from "pure math," is difficult for many people. This is like learning to play a guitar, I think—there is no substitute for practice. Please have a go at the exercises at the end of the chapter.

10. Why do you sometimes integrate from y_0 to y, and other times from $y = y_0$ to $y = y$?

• I use the first form when the variable of integration is y itself. I use the second form when I've performed a substitution, so that the variable of integration is some "u" or whatever. In the latter case, I want to remind myself that the limits on the integral are y values, not u values. This becomes especially important when the limits are numbers, rather than "y" and "y_0".

11. Setting up DEs is hard for me. Are there any steps you should *always* follow in that process?

• The variety of situations leading to DEs is sufficiently great that it seems unlikely there could be universal steps. However, with situations involving unsteady mass balances (our focus so far), the system I try to follow is to *work with mass*, and to set $m(t + h) \approx m(t) + h*$(sum of rates that increase m) - $h*$(sum of rates that decrease mass).

Then subtract $m(t)$ from both sides, divide through by h, and take the limit as $h \to 0$. This leads to an equation of the form $dm/dt =$ (sum of rates that increase m)-(sum of rates that decrease m). After you get more experience, or in some simple situations (as with the swallow), you can sometimes just write down the DE in a form like the last one. It depends on how complex the processes involved are.

Chapter 5

Further Topics in ODEs

This chapter considers two further but unrelated topics involving first-order ordinary differential equations: the diverse ways scientists in various fields describe and work with responses that approach a limit (an *asymptote*), then methods for solving certain DEs using *integrating factors*.

5.1 Asymptotic Behavior

We have seen numerous situations where some variable declines or increases asymptotically toward a limiting value, usually following a negative exponential. Examples are:

- Mercury increasing in a lake
- A swallow flying toward Capistrano
- Sediment settling from a river downstream from a source
- Decay of radioactive elements

It would be useful to have a measure of how long it takes such processes to come to completion, but they never quite do. For that reason, scientists in various fields have worked out ways to indicate how fast a response is "not quite getting there," i.e., how fast it approaches an asymptote. The terminology varies from field to field— e.g., engineers speak of time constants, limnologists of residence times and turnover rates, and radiation physicists and toxicologists of half-lives. All of these are measures of "asymptotic behavior."

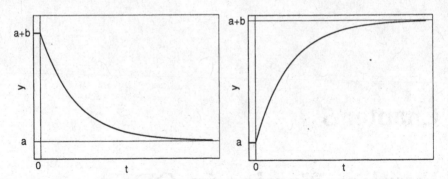

Figure 5.1: Processes described by differential equations with the solutions $y = a + b\exp(-kt)$ (left) and $y = a + b[1 - \exp(-kt)]$ (right).

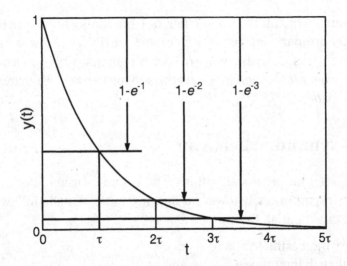

Figure 5.2: Progress of an exponential decay process as time passes through various numbers of time constants. The equation plotted is $y(t) = 1 \cdot \exp(-t/\tau)$.

Consider various processes described by $y = a + be^{-kt}$ or $y = a + b(1 - e^{-kt})$, and shown in Fig. 5.1. One way to describe such processes is via the general equation $y = y_f + (y_i - y_f)e^{-kt}$, where the subscripts i and f denote "initial" and "final," respectively. These processes both contain the form $y = e^{-kt}$, which, if τ is substituted for $1/k$, can be diagrammed as in Fig. 5.2.

Note that for each such process, $1 - e^{-1}$ of the total change occurs as t goes from 0 to τ (0 to $1/k$), where $e \approx 2.7183$. Thus e^{-1} is roughly

Table 5.1: Fractions of negative exponential processes completed as time progresses.

t	Fraction of total change completed
$1/k = \tau$	0.63
$2/k = 2\tau$	0.86
$3/k = 3\tau$	0.95
$4/k = 4\tau$	0.982
$5/k = 5\tau$	0.993
$6/k = 6\tau$	0.998
$\dots$	$\dots$
$10/k = 10\tau$	0.99995

1/3, and $1 - e^{-1}$ is roughly 2/3. Furthermore, a fractional change of $1 - e^{-2}$ occurs as t goes from 0 to 2τ, etc. From this, we can calculate the values in Table 5.1.

One point that arises here is not as well known as it should be— in any use of $\exp(x)$, the argument of the function (i.e., x) *must* be dimensionless (unit-free). One way to understand why is to think of the Maclaurin series that defines e^x. It is the sum of terms in x^0, x^1, x^2, and so on, and if x had units, this addition wouldn't be possible.

Time Constant

The quantity τ ($\stackrel{\text{def}}{=} 1/k$) is a useful measure related to the rate of exponential decay processes. It has units of time, and is called the *time constant*, especially in some fields of engineering.

Half-Life

When a substance has a half-life (H), that length of time is a constant multiple of the time constant. That is, if $y = y_0 e^{-rt} = y_0 e^{-t/\tau}$, then by definition, $y = y_0/2$ (y drops to half its original amount) when $t = H$. Thus $(y/y_0) = (1/2) = e^{-H/\tau}$, from which the half-life H is $(-\log 0.5)\tau$. The corresponding numerical values are $H \approx 0.693\tau$, or $\tau \approx 1.443H$. Both H and τ have dimensions of time.

In population biology and human demography, the similar concept of doubling time arises, but of course this applies to quantities that are *increasing* exponentially, at ever-increasing absolute rates.

Measures in Limnology

Suppose clean water flows at a rate q into a lake that initially contains substance X at a concentration C_0. Assume perfect mixing within the lake, outflow of water equal to inflow, and a constant lake volume V. Then the rate of change C in the lake water is described by

$$\frac{dC}{dt} = \frac{q}{V}(0 - C) = -\frac{q}{V}C, \qquad C(0) = C_0.$$

(The 0 here represents the concentration of X in the clean water flowing in.) The solution is

$$C(t) = C_0 e^{-(q/V)t} = C_0 e^{-t/\tau} \text{ with } \tau = V/q.$$

For example, if V is in m^3 and q in m^3 da^{-1}, then τ has units of days. In limnology, τ is known as the residence time, and its reciprocal (τ^{-1}) is called the turnover rate.

Note that, over the whole flushing process, the average length of time $(\bar{t}_w)$ that particular molecules of X stay in the lake is

$$\bar{t}_w = \lim_{t \to \infty} \frac{\int_0^t sC(s)V\,ds}{\int_0^t C(s)V\,ds}.$$

This is a *weighted mean*—the mean time of residence, weighted by the number of molecules present for each length of time. Analysis shows that $\bar{t}_w = \tau$, so τ is in fact the average length of time the molecules reside in the lake once the flushing processing starts.

5.2   Integrating Factors for Linear First-Order Differential Equations

When we dealt with first-order ODEs in Chapter 4, the most general solution method we considered was separation of variables. However, many important DEs remain that cannot be separated, the simplest being $y' = x + y$. Although solutions of many such equations are listed in tables, it is useful to be able to obtain solutions when tables and analytic software are not readily available. We will therefore consider one further method that is helpful for many (but still not all) linear, first-order ODEs.

The most general DE in this class is

$$\frac{dy}{dx} = f_1(x)y + f_2(x).$$

The Constant-Coefficient Case

We will begin with the *constant-coefficient* version of this equation, which we write as

$$\frac{dy}{dx} + ay = g(x), \tag{5.1}$$

where a is a constant. The idea is to find an *integrating factor I* that makes

$$\left(\frac{dy}{dx} + ay\right)I = g(x)I$$

easy to solve.

Although few of us would probably have figured this out for ourselves, it is easy to confirm (using the product rule for derivatives—p. 24) that

$$\frac{d}{dx}(e^{ax}y) = e^{ax}\frac{dy}{dx} + ae^{ax}y = e^{ax}\left(\frac{dy}{dx} + ay\right).$$

To confirm this, set $u(x) = e^{ax}$ and $w(x) = y$, and apply the product rule.

This suggests, or did to some creative mathematician many years ago, that we try $I = e^{ax}$ as an integrating factor. This yields

$$e^{ax}\left(\frac{dy}{dx} + ay\right) = e^{ax}g(x),$$

from which

$$\frac{d}{dx}(e^{ax}y) = e^{ax}g(x).$$

Now we can multiply through by dx and attempt to integrate both sides:

$$\int d(e^{ax}y) = \int e^{ax}g(x)dx.$$

Next we use the fact that $\int du = u - C$[1] for any u, so this expression becomes

$$e^{ax}y = \int e^{ax}g(x)dx + C,$$

which, because $1/e^{ax} = e^{-ax}$, we can write as

$$y = e^{-ax}\left[\int e^{ax}g(x)dx + C\right].$$

Be forewarned, and don't make the common error of writing this as

$$y = \left[e^{-ax}\int e^{ax}g(x)dx\right] + C,$$

which is not the same. (The constant has to be multiplied by e^{-ax}!) If we can perform the integration on the RHS, then we have our solution. Such integrals often fall to integration by parts, or better yet, can be found in tables like the one in Section 5.3, p. 131.

Consider an example. Suppose

$$y' = 1.3y + x^2, \text{ or } y' - 1.3y = x^2, \text{ with } y(0) = 3.$$

Here we have the form $y' + ay = g(x)$ (Eqn. 5.1) with $a = -1.3$ and $g(x) = x^2$. In our usual fashion, we first solve the symbolic form, so

$$y = e^{-ax}\left[\int e^{ax}x^2dx + C\right].$$

Any good table of integrals, or even the short table on p. 131, will show that

$$\int e^{kx}x^2dx = \frac{e^{kx}}{k^3}(k^2x^2 - 2kx + 2)$$

plus a constant, which can be added into the C already present. Thus, with k replaced by a,

$$y = e^{-ax}\left[\frac{e^{ax}}{a^3}(a^2x^2 - 2ax + 2) + C\right] \qquad (5.2)$$

$$= \frac{1}{a^3}(a^2x^2 - 2ax + 2) + Ce^{-ax}.$$

---

[1]The constant is arbitrary, so we make it negative simply because it is convenient to do so.

For $a = -1.3$, Eqn. 5.2 becomes

$$y = \frac{-1}{1.3^3}(1.3^2 x^2 + 2.6x + 2) + Ce^{1.3x}.$$

The value of C can be obtained from the IC, $y(0) = 3$, by setting

$$3 = -\frac{1}{1.3^3}(0 + 0 + 2) + C \cdot 1,$$

and solving for $C = 3 + 2/1.3^3$.

Clearly the success of this method depends on the integrability of $e^{ax}g(x)$.

Summary for the Constant-Coefficient Case, $y' + ay = g(x)$

The linear, first-order ODE with a constant coefficient multiplying y, i.e., $y' + ay = g(x)$, has the solution

$$y = e^{-ax}\left[\int e^{ax}g(x)dx + C\right],$$

which is useful whenever you can perform the integration. In practice, one just uses this formula—the derivation doesn't have to be repeated every time.

The Variable-Coefficient Case

The most general linear, first-order DE is the form

$$y' + f(x)y = g(x),$$

where the constant a in the situation just dealt with has been replaced by the function $f(x)$. The derivation of the integrating-factor method for this form is similar to that for the constant-coefficient case, but is more complex. We will skip the derivation, and simply state the result:

Summary for the Variable-Coefficient Case, $y' + f(x)y = g(x)$.

Given a linear, first-order ODE with a function of x multiplying y; i.e., $y' + f(x)y = g(x)$, set

$$F(x) \overset{\text{def}}{=} \int f(x)dx.$$

Now $e^{F(x)}$ is our integrating factor, and if we can do the integration analytically, the solution to the ODE is

$$y = e^{-F(x)}\left[\int e^{F(x)}g(x)dx + C\right].$$

When the necessary integrations can be performed, this method is a powerful one.

We demonstrate this method by solving

$$\frac{dy}{dx} + \left(\frac{b}{x}\right)y = cx^2,$$

for which $f(x) = b/x$ and $g(x) = cx^2$. Then

$$F = \int f(x)dx = \int \frac{b}{x}dx = b\log x = \log x^b.$$

Thus $e^F = \exp(\log x^b) = x^b$. This is our integrating factor. (It is a special case, though. Because F was a logarithm, the exponential function disappeared. Usually the exponential remains as part of the integrating factor.) Now

$$y = e^{-F}\left[\int e^F g(x)dx + C\right] = x^{-b}\left[\int x^b \cdot cx^2 dx + C\right].$$

Recall that for any x, u, and w, $x^u x^w = x^{u+w}$. (For example, $2^3 \cdot 2^2 = 2^{3+2} = 2^5$.) Therefore,

$$y = cx^{-b}\left[\int x^{b+2}dx + C\right] = cx^{-b}\left[\frac{x^{b+2+1}}{b+2+1} + C\right].$$

As usual, the value of C would be obtained from an initial condition.

Suppose, for a more specific example, that we wanted to solve

$$\frac{dy}{dx} + \frac{3}{x}y = 7x^2 \text{ with } y(2) = 4.$$

Then $b = 3$ and $c = 7$, so

$$y = 7x^{-3}\left(\frac{x^6}{6} + C\right).$$

From the IC (i.e., $y = 4$ when $x = 2$),

$$4 = 7 \cdot 2^{-3} \left(\frac{2^6}{6} + C \right),$$

so $C = -256/42$. Substituting this into the solution yields

$$y = \frac{1}{6} \left(7x^3 - 256x^{-3} \right).$$

You could, of course, check this solution using the methods of § 4.3, p. 94.

5.3 Integrals for the integrating-factor method

The integrating-factor method for solving linear differential equations typically leads, as a step along the way, to integrals of the form $\int e^{ax} f(x) \, dx$. It is therefore useful to have handy a table of some of the common integrals of that type. Here are a few. (In each case, C is a constant of integration that must be included in the solution.) The first three are for $f(x) = x$, x^2, and, more generally, x^n:

$$\int x e^{ax} \, dx = \frac{e^{ax}(ax - 1)}{a^2} + C,$$

$$\int x^2 e^{ax} \, dx = \frac{e^{ax}(a^2 x^2 - 2ax + 2)}{a^3} + C.$$

This pattern carries on to higher powers of x as follows:

$$\int x^n e^{ax} \, dx = \frac{x^n e^{ax}}{a} - \frac{n}{a} \int x^{n-1} e^{ax} \, dx.$$

You then apply this repeatedly until $n \to 0$.

A common $f(x)$ might be e^{bx}. Because $e^{ax} e^{bx} = e^{(a+b)x}$,

$$\int e^{ax} e^{bx} \, dx = \int e^{(a+b)x} \, dx = \frac{e^{(a+b)x}}{(a+b)} + C.$$

Sines and cosines appear in some applications. In simple form, we obtain

$$\int e^{ax} \sin bx \, dx = \frac{e^{ax}}{a^2 + b^2} (a \sin bx - b \cos bx) + C;$$

and

$$\int e^{ax} \cos bx \, dx = \frac{e^{ax}}{a^2 + b^2} (a \cos bx + b \sin bx) + C.$$

If the sine or cosine function starts at an x value other than zero (i.e., there is a *phase shift* represented by ϕ), then,

$$\int e^{ax} \sin b(x+\phi) \, dx = \frac{e^{ax}}{b^2 + a^2} \left[b - 2b (\cos \theta)^2 + 2a \sin \theta \cos \theta \right] + C;$$

and

$$\int e^{ax} \cos b(x+\phi) \, dx = \frac{e^{ax}}{b^2 + a^2} \left[2b \sin \theta \cos \theta - a + 2a (\cos \theta)^2 \right] + C,$$

where $\theta = b(x + \phi)/2$.

Finally, don't forget that the integral of the sum of two functions is the sum of their integrals. This would mean, for example, that

$$\int e^{ax} [f(x) + g(x)] \, dx = \int e^{ax} f(x) \, dx + \int e^{ax} g(x) \, dx.$$

5.4 Exercises

Asymptotic Variation

1. Lead 210 is a radioactive isotope that is frequently used to date lake sediments at various depths in a sediment profile. The process is like that for carbon 14 dating, but because the half-lives of ^{210}Pb and ^{14}C differ, these isotopes allow dating over quite different time intervals. The half-lives are, respectively, about $H = 22$ years and $H = 5770$ years.

   Suppose that the smallest loss of radioactivity that can be detected accurately for these two elements is 0.1% of the radioactivity initially present. Then what would be the shortest period of time before present that could usefully be measured by each isotope? Also suppose that the radioactivity levels for each isotope could be accurately detected until they dropped to 0.1% of initial levels. What lengths of time before present would these levels represent? Do the useful data ranges for the two isotopes overlap?

   It is most efficient to solve this problem analytically for arbitrary half-life H and arbitrary fractional mass loss $m(t)/m_0$. Then you can just plug in the numbers to get the answers for the four individual cases. (That has the further advantage that your answers would apply to other isotopes, like cesium 137.)

2. A lake in Minnesota has a water volume of 2×10^7 m³. The average annual discharge into, and out of, the lake is 5×10^6 m³. The lake starts out clean, but at $t = 0$ the incoming stream becomes polluted with an average of 0.3 mg L^{-1} of arsenic, caused by an industrial process carried out in a factory that discharges effluent into the stream. This discharge continues for many years.

A. Derive and solve the differential equation to obtain an estimate of [As] in the lake over time. What assumptions do you have to make here?

B. Calculate the time constant, the residence time, the turnover rate, and the half-life for this process. Give units for each.

C. Suppose the lake has an irregular shoreline, with islands and large bays. As a result, some portions of the lake contain water that does not mix readily with the water in the main basin. Would this fact cause the time constant for the true average arsenic concentration to be larger or smaller than the value for a well-mixed lake? (Your answer here should be qualitative rather than quantitative.)

D. Do the concentrations involved (i.e. $C_0 = 0$, $C_{in} = 0.3$ mg L^{-1}) have any effect on the time constants and related parameters?

Reminder: For parts A and B, work analytically as long as you can, and put in numbers only at the very end.

3. A pond on a university campus contains a water volume $V = 50$ m³, and water flows through it at a rate $q = 20$ m³ da^{-1}. How long would it take for all but 1/3 of the water present to be flushed through?

4. Suppose some radioactive waste from a nuclear power plant had a half-life of $H = 20,000$ years. In helping to design a waste depository, you want to model the decay of a starting mass, m_0 kg of that material, using the equation $dm/dt = -km$.

A. What would be the symbolic and numerical values of k, and what would be its units?

B. What would be the symbolic solution of that differential equation (without numbers)?

C. What fraction of the original mass would remain after 1000 years?

5. Consider a lake that has a constant volume V [m^3] of well-mixed water, a constant inflow q [m^3 da^{-1}] of arsenic-free water, a constant outflow q (the *same* q) [m^3 da^{-1}] of the lake water, and an initial mass m_0 [g] of arsenic dissolved evenly throughout the lake water.

   Determine the half-life H of the arsenic in the lake water, in terms of V and q. (H is the time required for the arsenic mass in the water to decline to half its initial value.) Show your work and reasoning in detail. Any processes not mentioned explicitly here may be assumed negligible. Be sure to check units.

6. A certain toxic chemical has an effective half-life of four months in the human body, with the chemical being eliminated primarily in the urine.

   A. What would be the time constant for this elimination process?

   B. Write and solve the differential equation for the mass of the substance in the body, as it varies with time. The initial mass is $m(0) = m_0$. Use symbols, not numbers, for this part. Provide a unit check (and a discussion of what you conclude from it) for your final equation.

   C. How long would it take for 90% of the mass that is present at time $t = 0$ to be eliminated? Work in symbols, substituting numbers (with units) only at the very end.

Integrating Factors

7. Use an integrating factor to solve $y' = 0.3x + 0.2y$, given $y(0) = 0$. Check your solution.

8. The Streeter-Phelps equations (p. 144), derived in the 1930s or thereabouts, provide a first-cut model for the depletion of oxygen in a stream or lake caused by the presence of "BOD" (biochemical oxygen demand). The equations are:

$$B' = -kB; \quad B(0) = B_0$$

$$D' = -kB + r(D_s - D); \quad D(0) = D_0,$$

where B=BOD concentration [mg L^{-1}], D=dissolved oxygen concentration [mg L^{-1}], k=reaction rate constant [hr^{-1}], and r=reaeration rate constant [hr^{-1}].

Solve the first equation by separation (or by inspection), substitute that solution into the D' equation, and solve the latter using an integrating factor. Note that the "oxygen sag" modelled by these equations has *two* characteristic time constants, one from the decay process and one from the deaeration. In such instances the larger time constant (representing the slower process) determines the time response of the system as a whole.

9. Solve $y' = \cos x - 2y/x$, if $y(\pi) = 0$. It will help to know that

$$\int x^2 \cos x \, dx = 2x \cos x + (x^2 - 2) \sin x + \text{constant}.$$

10. A certain toxic chemical has an effective half-life of 6 months in the human body, with the chemical being eliminated primarily in the urine.

   a) What would be the time constant for this elimination process? Please provide a number, with units.

   b) Write and solve the differential equation for the mass of the substance in the body as it varies with time. Please work symbolically here.

   c) How long would it take for 99% of the mass that is present at some time $t = 0$ to be eliminated? Work in symbols, substituting numbers only at the very end.

5.5 Questions and Answers

1. What's the difference between 'solving a differential equation' and 'integrating?'

   • Solving a DE [i.e., finding the function $y(t)$ that satisfies the DE $y' = f(t, y)$, maybe with an initial condition] usually involves integrating, but usually also involves some algebra. Integration is sometimes done for other purposes, like finding means, areas under curves, areas of surfaces, volumes of solids, and probabilities. Thus integration is a more general process.

2. Is $\exp(ax)$ always the integrating factor for the constant-coefficient case?

   • For such cases, the integrating factor is always of that form, yes.

Chapter 6

Systems of Ordinary Differential Equations

6.1 ODEs for Multiple Response Variables

In the previous two chapters we've considered situations in which our fundamental knowledge tells us about the rate of change of one variable (y, say) with respect to another (often time t), rather than about a simple function relating the two. Such situations provide us with differential equations of the form $dy/dt = F(t, y)$, which can often be solved to yield direct functional relationships; e.g., $y = f(t)$.

Frequently, variables involved in differential equations do not change by themselves, but rather as part of a larger *system* of interacting variables. An obvious example is that of interacting populations in an ecosystem. We could represent an ecosystem with three trophic (feeding) levels symbolically using interacting populations of carrots (C), rabbits (R), and foxes (F)[1]. One simple model of such a system (a linear system with constant coefficients) might be

$$\frac{dC}{dt} = a_{11}C + a_{12}R + a_{13}F + f_c(t); \quad C(0) = C_0 \tag{6.1}$$

$$\frac{dR}{dt} = a_{21}C + a_{22}R + a_{23}F + f_r(t); \quad R(0) = R_0$$

---

[1]You might think of these as convenient nicknames for the biomasses [kg] of all plants, all herbivores, and all carnivores in an ecosystem being modelled.

$$\frac{dF}{dt} = a_{31}C + a_{32}R + a_{33}F + f_f(t); \quad F(0) = F_0.$$

This model states that each population changes, at any instant, at a rate that depends linearly on the current sizes of the three populations, as well as (possibly) on time-dependent factors represented by the f functions. The coefficient a_{22}, for example, would be related to the rabbits' birth rate minus death rate, not counting the direct contributions of the carrots and foxes. The value of a_{21} would probably be positive, reflecting the greater growth potential of the rabbit biomass when their vegetable food supply was plentiful. On the other hand, a_{23} would usually be negative because a larger fox population would tend to cause a decrease in the biomass of rabbits. Coefficients a_{13} and a_{31} would be small or zero, since the carrots and foxes have little *direct* effect on one other—their interdependence is largely indirect, with the rabbits as mediators. The function f_1 could represent variations in carrot production that are due to seasonal climatic changes (but independent of C, R, and F), and so on for the other f terms.

Linear models like these were studied intensively by some mathematical ecologists in the mid to late 1960s, but they seemed to offer little insight into real ecosystems and are no longer an active subject for research. Linear systems models have the advantage that, for some simple fs, they can be solved semi-analytically (Derrick and Grossman 1981). But they have the major disadvantage that in most instances the linear dependence of rates on state variables is too simplistic to represent realistic population interactions.

It is not easy to model something as complex as an ecosystem mathematically, but many have tried. A famous example, called the Lotka-Volterra equations after the two mathematicians who independently developed them, can be written (for example for the rabbits and foxes) as

$$\frac{dR}{dt} = a_{11}R - a_{12}RF$$

$$\frac{dF}{dt} = a_{21}RF - a_{22}F.$$

Here the terms in R and F account for the dependence of the rates of change of each population on their own population sizes, while the

RF terms suggest that the interactions are proportional to the number or biomass of rabbits around to be eaten by foxes times the number or biomass of foxes around to eat the rabbits. This seems more logical than the $a_{12}R$ term in the linear carrot equation (Eqns. 6.1), since the latter implies that 100 rabbits would have the same effect on the carrot rate of change whether there were 2 kg of carrots, or 2 million, available for the rabbits to eat.

While the Lotka-Volterra equations may not be perfectly descriptive of real population interactions either, they do seem more logical than the linear system, and of course they can be generalized to even more complicated (and, one hopes, more realistic) forms (Haefner 2005). Unlike the linear system, however, they usually cannot be solved analytically. If one wants to plot their solutions to see their full behavior, they must be solved numerically (Chapter 7).

A great many environmental phenomena can be usefully modelled using systems of ODEs, but we will consider just one more example for now. (Additional applications are provided in the exercises.) In deriving the DE for mercury concentration in the lake (p. 83), we assumed a constant and equal streamflow into and out of the lake. We also assumed that evaporation of water was negligible, and that the mercury concentration in the incoming water was constant. Let us now add some greater realism and generality to the situation. We allow for

- the source stream to have a discharge and a mercury concentration that vary with time. Call these $q_i(t)$ and $c_i(t)$ respectively.

- the discharge of the outflow stream to depend on the water level (and hence the volume) of water in the lake. Thus $q_o = q_o(V)$, where now V varies rather than remaining constant.

- evaporation of water from the lake surface at a time-dependent rate $E(t)$. Although some evaporation of mercury would likely also occur, we will neglect that here.

- sedimentation processes to remove mercury from the water column. As a reasonable first approximation, this would occur at a rate proportional to the mass $m(t)$ of Hg present in the water column at any given time. Let k [da^{-1}] denote the proportionality constant for this process.

To account for all these phenomena, we need differential equations for both lake water volume, V, and mercury mass, m, because both are changing with time. To derive these DEs, we write approximate mass-balance equations to represent the changes of V and m from their values at any time t to some later time $t + \Delta t$:

$$V(t+\Delta t) \approx V(t) + \Delta t(q_i - q_o - E) \qquad [\text{m}^3] \approx [\text{m}^3] + [\text{da m}^3/\text{da}]$$

$$m(t+\Delta t) \approx m(t) + \Delta t(q_i c_i - q_o c - km) \qquad [\text{mg}] \approx [\text{mg}] + [\text{da mg}/\text{da}]$$

If we now subtract the $V(t)$ and $m(t)$ terms from both sides of their respective equations, divide through by Δt, and take limits as Δt goes to zero, we obtain

$$\frac{dV}{dt} = q_i - q_o - E, \text{ so } V' = q_i(t) - q_o(V) - E(t) = f_1(V, t) \quad (6.2)$$

and

$$\frac{dm}{dt} = q_i c_i - q_o c - km = f_2(c, t).$$

It is inconvenient to have the left and right sides of the latter equation in terms of different variables, m and c. The easiest way to deal with this is to substitute m/V for c in the RHS to obtain:

$$\frac{dm}{dt} = q_i c_i - \frac{q_o}{V(t)} m - km.$$

Thus, $m' = f(m, V, t)$. Together with Eqn (6.2), this provides a system of two first-order ODEs.

Given that $q_i(t)$, $q_o(V)$, $E(t)$, and $c_i(t)$ are not likely to be simple functions of time, it is almost certain that this system of equations (for V' and m') could not be solved analytically, and we would have to solve them numerically. Note that for any particular time at which you knew m and V, you could divide m by V to obtain the corresponding mercury concentration c. Note also that converting the m' equation to a c' equation would be complicated. Because $m = cV$ and V varies with time,

$$\frac{dm}{dt} = V\frac{dc}{dt} + c\frac{dV}{dt}, \text{ from which } \frac{dc}{dt} = \frac{1}{V}\left(\frac{dm}{dt} - c\frac{dV}{dt}\right).$$

This combination of derivatives is messy to work with.

In any case, it is now clear that there are important DEs, and systems of DEs, that can only be solved numerically, and that is the subject of the next chapter.

6.2   Exercises

Background for Exercises 1 and 2

Suppose a group of scientists has developed a model that estimates deposition rates of PCBs from the atmosphere to land in the
northern U.S. They believe it is reasonably accurate over the entire
period of industrial use of PCBs, from 1935 on, but they wish to test
it. They go to a relatively pristine northern lake, core the sediments,
and sample the PCB content of those sediments at various depths
that correspond to dates that they can estimate adequately.

Along with their deposition model, they need to model the PCB
content of the lake water as a function of time. They wish to account
for the following sources and losses of PCBs to and from the water in
the lake:

• There is one inlet stream, with the water flow rate of $q_{in}(t)$
[m^3 da^{-1}], which carries with it a concentration $C_{in}(t)$ [pg m^{-3}] of
PCBs. (One pg, or picogram, is 10^{-12} gram.)

• The atmospheric deposition rate of PCBs to the entire lake surface
is calculated from their model to be $D_a(t)$ [ng da^{-1}]. (One ng, or
nanogram, is 10^{-9} gram.)

• There is an outlet stream with a water flow rate of $q_{out}(t)$ [m^3 da^{-1}],
which carries PCBs at the lake water concentration (assuming good
mixing throughout the lake basin).

• Water evaporates from the lake at a rate of $E(t)$ [m^3 da^{-1}]. PCB loss
by evaporation is assumed negligible.

• PCBs enter the sediments at a rate proportional to the total amount
in the lake water. The instantaneous PCB sedimentation rate is a
proportion p_1 per year of the amount in the lake water, or $p_2 = p_1/365.25$ per day.

• All other potential PCB sources and losses are assumed to be negligible. Also, all isomers of PCB are assumed to be identical for the
purpose of these models, so the PCB amounts discussed are totals
over all isomers. Initially, at the end of 1934, the lake contains a
volume of V_0 [m^3] of water that is free of PCBs.

1. Derive two differential equations, such that the solution of the first
   would be the volume $V(t)$ of water in the lake at any number of

days (t) after 1/1/1935, and the solution of the second would be
the mass $B(t)$ [specify your units] of PCBs in the entire volume of
lake water. Write these equations in terms of the variables specified above.

2. Now make the following simplifying assumptions:

    A. $q_{in}(t) = q_{out}(t) = q_o$, a constant.

    B. C_{in} increases linearly with time; i.e., $C_{in} = \alpha t$

    C. D_a also increases linearly with time; i.e., $D_a = \beta t$.

    D. Evaporation of water (as well as PCBs) is zero.

    Then, obtain analytic solutions to the resulting simplified equa-
    tions, using separation, tables, or whatever is needed to do so.

3. Consider two small islands in the South Pacific that are connected
by a narrow isthmus, as shown in Fig. 6.1.

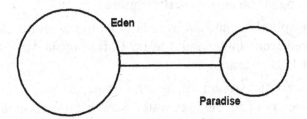

Figure 6.1: Two islands connected by an isthmus.

Captain Cook stops briefly at Eden, and a few rats escape from his
ship to that island. Thereafter the rat population density on Eden
grows logistically (similarly to Eqn. 1.5, p. 9) toward an areal carrying capacity of K_E [rats m^{-2}] with specific growth rate r_E [da^{-1}].
Eden has an area of A_E and Paradise of A_P, both in m^2.

As the population grows (assume uniformly over the whole area
of each island) the rats quickly discover the isthmus, and begin to
move to Paradise where the areal carrying capacity is K_P and the
specific growth rate is r_P (same units as for Eden).

That inter-island migration from E to P occurs at a rate proportional to the difference in rat *densities* on the two islands; i.e., to
$(\rho_E - \rho_P)$. Note that if ρ_P exceeds ρ_E at any time, the net migration

would then be toward Eden—the equation takes care of that automatically, as long as you give the migration term the right sign.

Your task is to derive two first-order ODEs, one for the rate of change of ρ_E and the other for the rate of change of ρ_P. (You will likely want to start with "mass balances" of rat *numbers* rather than of rat densities, however.) The initial conditions are $\rho_E(0) = \rho_0$ and $\rho_P(0) = 0$. You need not attempt to solve these equations.

Hint: Note that if K is the carrying capacity of an island in rats per square meter, and if A is the island's area, then KA is the carrying capacity in terms of rat numbers. (Check the units.)

4. Limnologists studying phosphorus cycling in a lake decide to model the P as cycling among three compartments—that in inorganic form (I), that in living biomass (L), and that in non-living organic matter (N). The mass in each compartment is expressed in units of mmol (millimoles) of P. The volume of water in the lake is V [L]. Suppose that transfers among the compartments occur as follows. (All rates are instantaneous, and all are in units of μmol da^{-1}.)

• From living biomass to non-living organic matter (by death and excretion), at a rate proportional to the mass of P in living biomass. (This is an example of a "donor-controlled" process).

• From non-living organic matter to inorganic P (by decomposition), at a rate proportional to the mass of P in non-living organic matter that is available to decompose.

• From inorganic P to living biomass at a rate proportional to the product of the masses of P present in those two compartments.

A. Derive the system of ODEs that describe these transfers. Leave these in terms of P masses in each compartment, not P concentrations.

B. What must the units be for the proportionality constants for each of those processes?

C. Suppose the proportionality constants for the three processes named above have numerical values of 0.22, 0.044, and 2.2 respectively. Also suppose that at some instant ($t = t_1$, say), $I = 0.12$, $L = 0.24$, and $N = 1.0$ mmol. Determine the numerical value of the net rate at which each compartment would be gaining or losing P at that instant. State for each compartment whether it is losing mass, gaining mass, or in equilibrium (unchanging). Also state the units of each rate.

You'll probably find it useful to draw a diagram showing the three compartments, with arrows representing the flows among them. This is optional, however.

5. Consider the *SIR* model (Kermack & McKendrick 1927) for a disease epidemic, using differential equations to connect three subgroups of a population:
   - Susceptible people, S
   - Infected people, I
   - Recovered, dead or otherwise immune people, R

   Suppose the rate at which susceptible people become infected is proportional to the product of the number of "susceptibles" times the number currently infected, and thus available to pass on the disease to the susceptible people. Suppose too that a fraction f of the people who are ill (infected) recover (or die) each day. Finally, suppose that once people recover (or die), they can't pass on the disease to others, and that they are immune to reinfection.

   Derive the differential equations that represent these assumptions. Be sure to specify some appropriate initial conditions.

6. Derive the Streeter-Phelps equations, which describe the sag of dissolved O_2 concentration D [mmol L^{-1}] in a stream below a source of "biochemical oxygen demanding" wastes ("BOD)," with concentration B [mmol L^{-1}].

   The two equations are based on these assumptions:

   - BOD mass is measured in "oxygen-equivalent" units, i.e., in units of the mass of O_2 that would be consumed in oxidizing it.

   - In any given slug (fixed mass) of water, BOD is lost at a rate proportional to the mass of BOD available to be oxidized as the slug moves downstream.

   - From the definition of the units of BOD, O_2 is also lost at a rate proportional to how much BOD is available to be oxidized.

- The stream regains dissolved oxygen by reaeration as it flows along with its surface in contact with the air above it. That process is modelled thus:

A. Water becomes saturated with O_2 at a temperature-dependent concentration D_s mmol L^{-1}. (Assume temperature is constant, and that D_s is known.)

B. The reaeration at any instant takes place at a rate proportional to how far the present concentration of DO lies below D_s. (The proportionality coefficient depends on the depth and turbulence of the water.)

7. Lipid-soluble chemicals that enter the human body are sometimes modelled as partitioning between a "blood" compartment and a "fatty-tissue" compartment, and we'll do that here. Suppose a factory worker takes up a substance S that she works with at a uniform rate[2] of U mg da^{-1}. As S circulates in her blood, a fraction f per day of what is in her blood is removed by her liver and kidneys. Suppose her blood has mass B kg and her fatty tissues available for exchanging S with the blood have mass F kg. At any instant, S exchanges between the blood compartment and the fat compartment at a rate proportional to the difference between its concentration in the blood (in mg kg^{-1}) and p (a dimensionless partition coefficient) times its concentration in the fat (in the same units).

A. Derive the system of ordinary differential equations that describe the net rates of change of S mass in each compartment. You may assume initial conditions of zero for both compartments, but that is irrelevant to the task of *deriving* the DEs.

B. Convert that to a system of equations for rates of change of S concentration, assuming that B and F are constant over some period of interest to a toxicologist.

These equations have an analytic solution, but we haven't studied how to obtain it.

8. The following scenario involving indoor air pollution involves two rooms of a house. The two rooms are connected by an open door,

---

[2]In reality, this rate would vary between times she's at work and times she is not, but over the long run, we can work with the average. Let's hope U is a small number.

but are isolated from other rooms. The rooms each have ceiling fans, so you may assume that within each room the air is well mixed.

In Room 1, on the north side of the house, a new carpet has just been laid. An organic solvent, a component of the cement used to lay the carpet, is evaporating into the air. The evaporation rate per m^2 of floor (in $\mu g\ m^{-2}\ s^{-1}$) is proportional to the difference between the concentration C_0 ($\mu g\ m^{-3}$) of the solvent in the air around the carpet fibers and the concentration of the solvent in the Room 1 air. The proportionality constant here is k_1 ($\mu g\ m^{-2}\ s^{-1}$) per ($\mu g\ m^{-3}$), which reduces to $k_1\ m\ s^{-1}$.

A breeze from the north brings air into Room 1 at a velocity V_1 ($m\ s^{-1}$) through a window with an opening of A_1 (m^2). That same amount of air ultimately flows through the doorway into Room 2, and out an open window on the south side of that room.

Room 2 has no new sources of the solvent, and we'll neglect any adsorption to surfaces there. However, a fraction f of the solvent in the air of Room 2 is destroyed by photolysis each second, a process driven by sunlight coming through the windows into that room.

The volumes of air in the two rooms are W_1 and W_2 m^3, respectively, and the surface areas of the two floors are S_1 and S_2 m^2.

Derive the differential equations for the concentrations $C_1(t)$ and $C_2(0)$ of solvent in the two rooms, if V_1 is a function of time. Assume initial concentrations of C_{10} and C_{20}, respectively. If $V_1(t)$ were known, we could solve this system numerically, but we leave that for another day.

Chapter 7

Numerical Solution of Ordinary Differential Equations

As noted in the last two chapters, many ODEs and systems of ODEs cannot be solved analytically. When we want solutions to such equations, we need to turn to numerical methods, which we now take up. We begin with a simple DE that can also be solved analytically by separation of variables[1]; i.e.,

$$\frac{dy}{dx} = xy, \ y(1) = 1,$$

This will allow comparison of the approximate numerical solution with the exact analytic solution, which is $y = \exp[(x^2 - 1)/2]$.

7.1 Euler's Method

We will begin with a very simple numerical process, called Euler's[2] method, that is too inaccurate for practical use but that demonstrates the basic pull-yourself-up-by-the-bootstraps idea underlying all such methods. (Two other reasons for studying this essentially useless method are [1] that some more accurate methods make use of it, and

---

[1] In practice, we therefore *should* obtain the analytic solution—the only reason for using this example here is to allow us to check the accuracy of the method.

[2] The name is pronounced "Oiler."

[2] to help you recognize when others use it indiscriminately. In the latter situation, you may be able to point the users to better methods).

In Euler's method, we replace the differential equation with a corresponding finite-difference approximation. Specifically, we use a *forward difference* (p. 33). It might be said (with apologies) that we "untake the limit" that gave us the derivative in the first place. Thus

$$\frac{dy}{dx} = xy \rightarrow \frac{\Delta y}{\Delta x} \approx xy.$$

The latter quantity is, of course, the slope of the curve at the point x, y. Because

$$\Delta y \overset{\text{def}}{=} y(x + \Delta x) - y(x) \approx xy\Delta x,$$

we have

$$y(x + \Delta x) \approx y(x) + \Delta x(xy).$$

If we now set $h \overset{\text{def}}{=} \Delta x = 0.1$, we can use this formula repeatedly, starting from the known IC, to get

$y(1) = 1$ (the IC)

$y(1.1) \approx y(1) + \Delta x \cdot x \cdot y = y(1) + 0.1(1)(1) = 1 + 0.1 = 1.1$

$y(1.2) \approx y(1.1) + 0.1(1.1)(1.1) = 1.1 + 0.121 = 1.221$

$y(1.3) \approx y(1.2) + 0.1(1.2)(1.221) = 1.221 + 0.14652$

$= 1.36752$, etc.

Notice the pattern—starting from a known point on the solution, we calculate the slope along which we should move away from that point. We move a little way along that slope to a new point. There we calculate a new slope, and move along that to yet another point, and so on. So, the pattern is point → slope → point → slope → .... (This is a little like stepping along the arrows on a direction field.) The problem is, of course, that we should move along each new slope for only an infinitesimal distance, but instead we take a comparatively gargantuan step of length Δx. (*Any* finite Δx is huge, compared with zero.)

Corresponding values from the *analytic* solution are $y(1) = 1$, $y(1.1) = 1.11071$, $y(1.2) = 1.246083$, and $y(1.3) = 1.41199$. Although the numerical solution is rising in the same general sort of way as the analytic, the former is clearly falling behind the latter, with ever-increasing error.

Now let us use h in place of Δx, and state Euler's method more generally. At x_0, the DE is

$$\left.\frac{dy}{dx}\right|_0 = f(x_0, y_0),$$

i.e., the slope is the value calculated from the slope function f based on the starting values of x and y. Replacing that with its finite-difference approximation yields

$$\left.\frac{\Delta y}{\Delta x}\right|_0 = \left.\frac{\Delta y}{h}\right|_0 \approx f(x_0, y_0).$$

Because Δy stands for the difference $y(x_0 + h) - y(x_0)$, then

$$\Delta y = y(x_0 + h) - y(x_0) \approx hf(x_0, y_0).$$

After moving the $y(x_0)$ term to the right side, we see that the general form can be written as

$$y_{\text{new}} \approx y(x_{\text{old}}) + hf(x_{\text{old}}, y_{\text{old}}).$$

We now apply this computation repeatedly, as with

$$y_2 = y(x_2) = y(x_1 + h) \approx y(x_1) + hf(x_1, y_1), \text{ etc.}$$

Consider another example; let

$$\frac{dy}{dx} = \frac{y}{4}, \text{ with } y = 1 \text{ when } x = 0.$$

We can compare the analytic solution ($y = 1 \cdot e^{x/4}$) with the Euler-method solution obtained using $h = 1$:

x	0	1.0	2.0	3.0	4.0	5.0	6.0
y_{true}	1	1.284	1.649	2.117	2.718	3.490	4.482
y_{Euler}	1	1.250	1.562	1.953	2.441	3.052	3.815

These results are also shown in Fig. 7.1, p. 150. Note the following points on the inaccuracy of Euler's method:

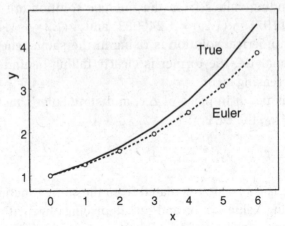

Figure 7.1: Comparison of the Euler-method solution with the true solution of $y' = y/4$, with $y(0) = 1$.)

- The slope is calculated at the left end of each Δx interval. However, the true slope should increase continuously as x increases.

- Thus at each step, y increases less than it should, and this would occur even if the starting value of y for that step were correct.

- To make matters worse, once beyond the first step, the slopes are low for a second reason as well—each slope is calculated as $y/4$, but if y is already too low from errors in previous steps, this will reduce the slope even further below what it should be.

For this reason, I emphasize that Euler's method should not be used in practice. Its main value is its simplicity, which helps to show the general concept of how numerical solution of DEs works. (See footnote, p. 148.)

Before moving on to a useful solution method, we note that Euler's method (like other methods) can be generalized to *systems* of ODEs. The same pattern, of $y \rightarrow y' \rightarrow y \rightarrow y' \rightarrow \ldots$ holds, except now each y represents a *vector* of state variables, one for each DE in the system. For example, consider again the Lotka-Volterra equations, which, as we have seen, cannot be solved analytically:

$$R' = aR - bRF = \text{"rabbit slope"}, \ R(0) = R_0$$

$$F' = cRF - dF = \text{"fox slope"}, \ F(0) = F_0.$$

If we used a time step of h years, Euler's method would proceed thus:

$$R_1 \stackrel{\text{def}}{=} R(0 + h) \approx R_0 + h(aR_0 - bR_0F_0)$$

$$F_1 \stackrel{\text{def}}{=} F(0 + h) \approx F_0 + h(cR_0F_0 - dF_0)$$

Thus, the *slopes* for both state variables are calculated from the *values* of both at $t = 0$. Further steps follow the same pattern:

$$R_2 \stackrel{\text{def}}{=} R(2h) \approx R_1 + h(aR_1 - bR_1F_1),$$

$$F_2 \stackrel{\text{def}}{=} F(2h) \approx F_1 + h(cR_1F_1 - dF_1),$$

and so on. Again, however, Euler's method is only illustrative.

We now turn to the fourth-order Runge-Kutta method, a real workhorse in the stable of methods for solving ODEs numerically. We begin by applying it to a single, simple DE.

7.2 The Runge-Kutta Method

Although there are *many* Runge-Kutta methods for integrating differential equations (Press et al. 1992), we will cover only one, a simple and common fourth-order method. We will apply it first to the simple, first-order ODE,

$$y' = \frac{dy}{dx} = f(x, y) \text{ with IC } (x_0, y_0).$$

Our goal will be to estimate the value of y at $x = x_0$, $x_0 + h$, $x_0 + 2h$, $x_0 + 3h$, etc. As with Euler's method, we use our estimate of the IC to estimate $y(x_0 + h)$, then use that point to estimate $y(x_0 + 2h)$, and so on. Let us concentrate on the first step away from the IC.

With Euler's method, we calculated a single slope, and moved from P_0 (point 0) to P_1 (point 1) along that slope. With our R-K method, we calculate four different slopes, and then move from P_0 to P_1 along a line with a slope that is a weighted mean of the four. (The idea is similar to Simpson's rule, which you may recall estimates the value of an integral based on a weighted mean of three function values—here we use a final slope estimate that is the weighted mean of four trial slopes.)

Here is the procedure:

1. Define point $P_0 \overset{\text{def}}{=} (x_0, y_0)$; i.e., the initial condition, and slope $S_a \overset{\text{def}}{=} f(P_0) = f(x_0, y_0)^3$. That is, the first slope is calculated from the x and y values at the known point from which we are moving. This first slope is the same one used in an Euler-method step.

2. "Pull yourself up by the bootstraps," and extrapolate along slope S_a from point P_0 to a point P_a that lies a distance $h/2$ to the right of P_0 in the x direction. Thus $P_a = (x_0 + h/2, \; y_0 + hS_a/2)$. (In this sub-step, we go only half as far as we would in an Euler-method step.)

3. Let $S_b \overset{\text{def}}{=} f(P_a)$, and again extrapolate over $\Delta x = h/2$ from P_0, but along this *new* slope. This takes us to $P_b \overset{\text{def}}{=} (x + h/2, \; y + hS_b/2)$.

4. Let $S_c \overset{\text{def}}{=} f(P_b)$. Now extrapolate again from P_0, but this time go all the way to $x + h$. Call this new point P_c ($\overset{\text{def}}{=} x + h, \; y + hS_c$).

5. Now we are at an *estimate* of the final point of this step, but it won't be our final value. Instead, we use this estimated point at the right-hand end of our interval to estimate a fourth slope, $S_d \overset{\text{def}}{=} f(P_c)$, at the right-hand end.

6. We now have four estimates—S_a, S_b, S_c, and S_d—of the overall slope from (x_0, y_0) to $(x_1, y(x_1))$. One estimate lies at the left end of the interval, two lie in the center, and one lies at the right end. We use an argument similar to the one that justifies using the central-difference formula rather than the forward-difference form for numerical differentiation—a slope near the center of an interval should be a better estimate of the overall slope of a function over that interval than a slope near the ends would be. For this reason, we average the four slope estimates, but place greater weight on the two mid-interval slopes. Thus we set

$$\overline{S} \overset{\text{def}}{=} \frac{1 \cdot S_a + 2 \cdot S_b + 2 \cdot S_c + 1 \cdot S_d}{1 + 2 + 2 + 1}.$$

(The 1:2:2:1 weights are not arbitrary, but have been derived by

---

[3]Often the independent (x) variable does not appear on the right-hand side of the differential equation. When that is true, then the slope values that we calculate depend only on the current value of "y", and not on the current value of "x". The instructions here, and the first example, deal with the more general case where the slope function depends on both variables, however.

Runge and Kutta, to make the method a fourth-order method. This point is discussed below.)

7. Finally, we extrapolate a distance h along the slope $\overline{S}$ from P_0 to a new point $P_1 \overset{\text{def}}{=} [x + h, y(x + h)]$. This is the best estimate we will get of the value of y on the vertical axis corresponding to $x + h$ on the horizontal axis.

To continue on, we treat our new point (P_1) as a new IC, take another step (with four slopes), and continue stepping along in this way until we get to our goal.

So much for the formulas—it's time for a demonstration. As we did with Euler's method, we will solve a DE with known analytic solution so we can check our result:

$$y' = \frac{dy}{dx} = f(x, y) = x + \frac{y}{3}; \; x_0 = 0, \; y_0 = 0.6.$$

We will use Runge-Kutta with $h = 1$ to step to $x = 1$, a point we'll call (x_1, y_1). The calculations proceed as shown in the table below, which is laid out in spreadsheet-like form. The sequence of steps is illustrated graphically in Fig. 7.2, p. 154.

	x	y	S
P_0	0	0.6	
S_a			$0 + 0.6/3 = 0.2$[a]
P_a	$0 + 1/2 = 0.5$	$0.6 + (1/2) \cdot 0.2 = 0.7$	
S_b			$0.5 + 0.7/3 = 0.7\overline{3}$[b]
P_b	$0 + 1/2 = 0.5$	$0.6 + (1/2) \cdot 0.7\overline{3} = 0.9\overline{6}$	
S_c			$0.5 + \frac{0.9\overline{6}}{3} = 0.8\overline{2}$[c]
P_c	$0 + 1 = 1$	$0.6 + 1 \cdot 0.8\overline{2}) = 1.4\overline{2}$	
S_d			$1 + \frac{1.4\overline{2}}{3} = 1.4\overline{740}$[d]

Note that what might be called 'P_d' is never used. Now we calculate the weighted mean of the four slope estimates just found; i.e.,

$$\overline{S} = \frac{0.2 + 2(0.7\overline{3}) + 2(0.8\overline{2}) + 1.4\overline{740}}{6} = 0.797530864,$$

---

[a] $= f(P_0)$
[b] $= f(P_a)$
[c] $= f(P_b)$
[d] $= f(P_c)$

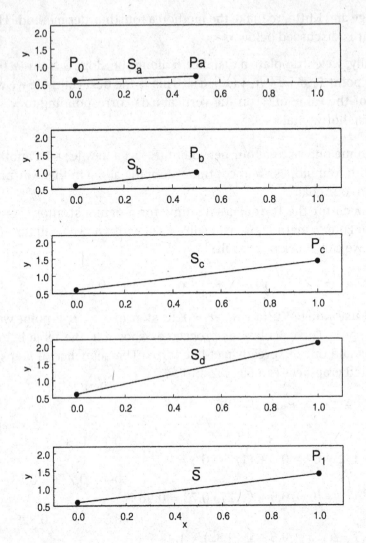

Figure 7.2: Illustration of points and slopes used in one Runge-Kutta step. Refer to the calculations on p. 153.

and move along that slope a distance h from P_0 to P_1. Thus

$$P_1 \stackrel{\text{def}}{=} (x_0 + h, y_0 + h\bar{S}) = (0 + 1, 0.6 + 1 \cdot 0.797530864) \qquad (7.1)$$

$$= (1, 1.397530864) = (x_1, y_1).$$

This is our best estimate of $y(x + h)$; i.e., $y(1) = 1.397530864$. It is important not to round off here, since this is our best estimate of

y at $x = 1$, and we need the best estimate so the slopes to the next point can be estimated as accurately as possible. Given our new P_1, we can now step on to the point P_2 at $x = 2$ by a similar series of steps, and then continue on for as many steps as necessary.

To check our approximate value of $y(1)$, we find that the second equation in the solution table on p. 107 yields $y = 9.6\exp(x/3) - 3x - 9$ as the analytic solution to our DE. Thus our estimate differs from the exact value of $y(1) = 1.397879280$ by about -0.00035, giving a relative error of -0.0249%. This may be "good enough for government work", but of course it represents a single step, and the error could accumulate if we integrated over many steps. We could probably improve the accuracy by using more, smaller steps.

Note that to go from $x = 0$ to $x = 1$ above, we evaluated $y' = f(\dot{x}, y)$ four times. An equivalent number of evaluations of the slope function would be required for four Euler method steps with $h = 0.25$:

$$y' = x + \frac{y}{3}; \quad x_0 = 0; \quad y_0 = 0.6; \quad h = 0.25$$

$$y(0.25) \approx 0.6 + 0.25\left(0 + \frac{0.6}{3}\right) = 0.65$$

$$y(0.5) \approx 0.65 + 0.25\left(0.25 + \frac{0.65}{3}\right) = 0.7\overline{6}$$

$$y(0.75) \approx 0.7\overline{6} + 0.25\left(0.5 + \frac{0.7\overline{6}}{3}\right) = 0.9\overline{5}$$

$$y(1) \approx 0.9\overline{5} + 0.25\left(0.75 + \frac{0.9\overline{5}}{3}\right) = 1.222685185.$$

We can now compare the value of $y(1)$ estimated by the two numerical methods with the analytic value:

Euler	$y(1) \approx 1.[2\ldots]$
Analytic	$y(1) = 1.397879280$
R-K	$y(1) \approx 1.397[5\ldots]$

The two estimates have required the same number of evaluations of the y' function (a common measure of the "cost" of a numerical method), but the Runge-Kutta integration has provided considerably greater accuracy. This results from the difference in order of the two methods. Euler's method (which is of first order) is exact only for a DE of the form y' =constant, while our fourth-order Runge-Kutta scheme would give exact answers (regardless of the size of h) for any DE of form $y' = a + bx + cx^2 + dx^3$. The *solution* of the latter equation[4], of form $b_0 + b_1x + b_2x^2 + b_3x^3 + b_4x^4$, is, of course, a fourth-order polynomial.

Runge-Kutta for an ODE System

We now look at an example of using the Runge-Kutta method to solve a system of two first-order ODEs. Earlier (p. 134) we considered the Streeter-Phelps equations, a classic model for the loss and subsequent replacement of oxygen from water downstream from a discharge of organic wastes into a stream. Those equations are fundamentally illogical in one sense, however—if the oxygen in the water were entirely depleted, they allow the oxidation to continue anyway. Here we will modify the equations to the form

$$B' = -kBD; \quad B(0) = B_0$$

$$D' = -kBD + r(D_s - D); \quad D(0) = D_0, \qquad (7.2)$$

where

B=BOD concentration [mg L^{-1}]

D=dissolved oxygen concentration [mg L^{-1}]

D_s= saturation DO concentration at the stream temperature [mg L^{-1}]

k=reaction rate constant [L mg^{-1} hr^{-1}] r=reaeration rate constant [hr^{-1}]

Although the original equations could be solved analytically, the present modified ones cannot. Let us solve them numerically for the case when

$B_0 = 100$ mg L^{-1}

$D_0 = 12$ mg L^{-1}

---

[4]That equation can be solved analytically, but it suggests approximating other more complicated equations by Taylor series

$D_s = 14$ mg L^{-1}

$k = 0.04/24$ L mg^{-1} hr^{-1}, and

$r = 0.5/24$ hr^{-1}

We will take one step with $h = 0.3$ hr (starting from an initial time of zero), to illustrate how the calculations work.

For our single illustrative step, the starting point is

$$P_0(t_0, B_0, D_0) = (0, 100, 12),$$

and from that we calculate the two starting slopes,

$$S_{Ba} = B'(P_0) = -kB_0D_0 = -2$$

$$S_{Da} = D'(P_0) = -kB_0D_0 + r(D_s - D_0) = -1.958333.$$

We step along those slopes for a time of $h/2$ to

$$P_a = [t_0 + h/2, B_0 + (h/2)(S_{Ba}), D_0 + (h/2)(S_{Da})]$$

$$= [0 + 0.15, 100 + 0.15(-2), 12 + 0.15(-1.958333)]$$

$$= [0.15, 99.7, 11.70625].$$

Now use that point to get the next slope estimates:

$$S_{Bb} = B'(P_a) = -1.945189$$

$$S_{Db} = D'(P_a) = -1.897402$$

and move along those slopes to the next point:

$$P_b = [t_0 + h/2, B_0 + (h/2)(S_{Bb}), D_0 + (h/2)(S_{Db})]$$

$$= [0 + 0.15, 100 + 0.15(-1.945189), 12 + 0.15(-1.897402)]$$

$$= [0.15, 99.708222, 11.71539].$$

Next calculate more slopes:

$$S_{Bc} = B'(P_b) = -1.946868$$

$$S_{Dc} = D'(P_b) = -1.899272.$$

Now determine the last temporary point:

$$P_c = [t_0 + h, B_0 + hS_{1c}, D_0 + hS_{2c}]$$

$$= [1 + 0.3, 100 + 0.3(-1.946868), 40 + 0.3(-1.899272)]$$

$$= [0.3, 99.41594, 11.430218]$$

and for the last set of intermediate slopes:

$$S_{Bd} = B'(P_c) = -1.89391$$

$$S_{Dd} = D'(P_c) = -1.840373.$$

Next, we calculate the weighted averages of the four slope estimates for each of the variables B and D:

$$\overline{S}_B = \frac{S_{Ba} + 2S_{Bb} + 2S_{Bc} + S_{Bd}}{6} = -1.946337$$

$$\overline{S}_D = \frac{S_{Da} + 2S_{Db} + 2S_{Dc} + S_{Dd}}{6} = -1.898676,$$

and finally step for a time h along those slopes to the best estimates we will obtain of the values of B and D at $t = t_0 + h$:

$$P_1 = [t_0 + h, B_0 + h\overline{S}_B, D_0 + h\overline{S}_D]$$

$$= [0 + 0.3, 100 + 0.3(-1.946337), 12 + 0.3(-1.898676)$$

$$= [0.3, 99.416099, 11.430397].$$

Thus, our best estimates for the values of the two response variables at $t = 0.3$ hr are

$$B(0.3) \approx 99.416099 \text{ and } D(0.3) \approx 11.430397.$$

We could now treat these values as a new starting point, and take a second R-K step to P_2 at $t = 0.6$ hr. In fact, we could repeat the process for as long as the solution remains of interest to us.

Note that neither the BOD nor the DO has declined much in this first 18 minute period. However, the BOD has declined by a greater net amount than has the DO, which is logical because some reaeration has taken place during the period. Note also that the four slope estimates were reasonably stable during the integration step, for both variables. This indicates that our time step h has been adequately small.

Admittedly it is not very satisfactory to consider this single time step, the purpose of which has been to illustrate the *process* of numerically integrating a system of differential equations. You could continue the solution over as many steps as you wished using a spreadsheet, if that were the best tool you had available. Better yet, use a tool like MATLAB as shown in the next section, if you have that available.

7.3 Solving ODEs Numerically with MATLAB

Solving one ODE, or a system of them, numerically using MATLAB®is a two-step process. First, one must create what MATLAB calls an *m function* that computes the slopes for the various *state variables*, given current values for the independent variable (e.g., time) and of the state variables. To do that for the modified Streeter-Phelps system (Eqn. 7.2), enter "edit" (without the quotes) at the MATLAB » prompt, and when an editor window pops up, enter the following five lines into that window (by typing them directly, or by cutting and pasting from some other source). Then in that window's menus, enter File | Save as, and save the file as "slopes.m".

```
function dydt = slopes(t,y)
% y is a column vector; y(1)=BOD, y(2)=DO
Ds=14; r=0.5/24; k=0.04/24;
dydt = zeros(2,1);
% That defines dydt as a vector the same length as y
dydt(1) = -k*y(1)*y(2);
dydt(2) = -k*y(1)*y(2)+r*(Ds-y(2));
```

The percent signs in the lines above tells MATLAB that the text to the right of them represents comments, to be ignored.

Next, we can calculate the numerical solution using a sophisticated variation of the Runge-Kutta method called ode45 in MATLAB. (Have a look at "help ode45.") To obtain the solution, enter [Time,Y]=ode45(@slopes,[0,500],[100 12]).
This computes a matrix of times and corresponding "y" values, with

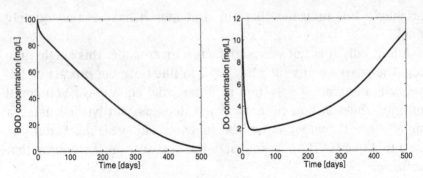

Figure 7.3: Solution curves for Runge-Kutta solution of the BOD-DO system model, as obtained from MATLAB. BOD, left; DO, right.

time running from 0 to 500 hours, and with initial conditions of 100 mg L^{-1} for BOD (y_1) and 12 mg L^{-1} for DO (y_2).

Finally, we can inspect our approximate solution by asking MAT-LAB to plot it. We could use a single plot, but because BOD starts out nearly ten times higher than DO, two plots work better. These lines do the work:

```
plot(Time,Y(:,1));
ylabel('BOD concentration [mg/L]');
xlabel('Time [days]')
% View the first plot (and save if desired).
plot(Time,Y(:,2));
ylabel('DO concentration [mg/L]');
xlabel('Time [days]')
```

The result is Fig. 7.3. Note that with the coefficients specified, the DO drops from 12 to about 2 mg L^{-1} rapidly, and then recovers more slowly by reaeration. The BOD also drops rapidly at first, but its loss then approaches a constant rate that is in a balance with the reaeration. The fish in this stream appear to be in some trouble!

7.4 Exercises

1. Consider the differential equation

$$\frac{dy}{dt} = 2.7t^{0.2}y^{0.1}, \text{ with } y(1) = 0.7.$$

As a baseline against which to compare numerical results, solve this equation analytically.

Next use Euler's method with a step size of $h = 0.1$ to estimate the value of y when $t = 1.4$. Plot the y values as you go along. Also plot the analytic solution for comparison. This exercise is designed to give you an intuitive feeling for what a numerical solution of a differential equation is. In real applications, you should avoid Euler's method because of its inferior accuracy relative to the Runge-Kutta method. Compare your result from this and the Runge-Kutta result to follow with the analytic solution.

Finally, use Runge-Kutta with $h = 0.2$ to estimate the value of y when $t = 1.4$. Plot the y values you obtain at $t = 1.2$ and $t = 1.4$ on the same graph as before. It will help to keep a summary table of the various t values, y values, and slopes that go into your solution.

2. Consider again the SIR model (p. 144) for a disease epidemic based on differential equations that connect three subgroups of a population; susceptible people, S; infected people, I; and recovered, dead or otherwise immune people, R.

   Now suppose the probability, and therefore the rate, of infection for people in the S group is proportional to the product of S times I. (Does that make sense to you?) Further, suppose a fixed fraction of infected people will recover per unit time. This model is sometimes called the *SIR* model.

   Next, suppose that at some early stage in a given epidemic, $S = 9900$ people, $I = 100$ people, and no one has yet recovered. Also, suppose that the proportionality constant for the S–I interaction is 4.2×10^{-5} person^{-1} day^{-1} and that 10% of infected people will recover per day.

   Using our Runge-Kutta method, take two one-day steps to estimate $S, I,$ and R two days after the time for which starting values were given. Keep a table of values of t, S, I, R, m_s, m_i and m_r, where the m values are slopes for the three variables.

3. The purpose of this exercise is to show you the relationships involved in changing the time units of differential equations. Let 1 year = 365 days. Then convert the differential equations from the

previous exercise so their time units are per year rather than per day. Finally, repeat the Runge-Kutta calculations using two steps with $h = (1/365)$ year and compare with the original calculations. Describe in words what happens.

4. As you have seen, the Lotka-Volterra equations for rabbit (R) and fox (F) populations are:

$$R' = aR - bRF; \quad R(0) = R_0 \tag{7.3}$$

$$F' = cRF - dF; \quad F(0) = F_0. \tag{7.4}$$

Consider a case where $a = 0.2$ year^{-1}, $b = 0.005$ fox^{-1} year^{-1}, $c = 0.00005$ rabbit^{-1} year^{-1}, $d = 0.04$ year^{-1}, $R_0 = 1000$ rabbits, and $F_0 = 20$ foxes.

Estimate the sizes of the rabbit and fox populations at $t = 0.2$ year (a little over two months after $t = 0$) using two Runge-Kutta steps of $h = 0.1$ year each. You will find this exercise easier if you keep a tidy table of values of t, R, F, s_r and s_f, where s_r values are "rabbit slopes" (dR/dt) and s_f values are "fox slopes" (dF/dt).

5. Perform one step (with $h = 0.1$ yr) of a Runge-Kutta solution of the modified Lotka-Volterra equations from pp. 17–18 of the notes:

$$\frac{dH}{dt} = r_H H \left(\frac{K_H - H}{K_H} \right) - aHC \text{ with } H = H_0 \text{ at } t = 0, \text{ and}$$

$$\frac{dC}{dt} = r_C C \left(\frac{bH - C}{bH} \right) + cHC \text{ with } C = C_0 \text{ at } t = 0.$$

For parameter values, use $H(0) = H_0 = 10,000$ [kg], $C(0) = C_0 = 1000$ [kg], $r_H = 0.14$ [yr^{-1}], $K_H = 2000$ [kg], $a = 0.002$ [kg yr^{-1}], $r_C = 0.06$ [yr^{-1}], $b = 0.1$ [-], and $c = 0.002$ [kg yr^{-1}].

Retain at least 5 digits in all calculations. (In a spreadsheet or in MATLAB, you would automatically be retaining more than that.)

6. Use MATLAB or other software to repeat Exercises 1–5, and check the results against what you found in your less-automated calculations.

7. As you know, the solution to $dN/dt = rN$ is $N(t) = e^{rt}$ when $N(0) = 1$. Thus, (A) $dN/dt = rN$, $N(0) = 1$ and (B) $dN/dt = r e^{rt}$, $N(0) = 1$ are equivalent mathematical statements, meaning

that (A) and (B) are differential equations that must have the same solutions.

Suppose you were to solve each of (A) and (B) numerically using Euler's method from $0 \le t \le 10$, using $h = 0.5$. Would the errors (the deviations from the true analytic solution) caused by using Euler's method be the same in both cases? If so, state why. If not, state which would result in the worse error, and why.

7.5 Questions and Answers

1. Does the Runge-Kutta method work for non-linear functions; i.e., those that aren't straight lines?

   • Yes. In fact, if your function were a straight line, the slope would be a constant, and then there would be no need for a numerical method. In that case you could always get an analytic solution. In the notes, I point out that the 1:2:2:1 weights for averaging the four slope estimates were chosen to make this method exact for any DE whose solution is a fourth-order polynomial in x (or t). That is, if you differentiated the polynomial $y = a + bt + ct^2 + dt^3 + et^4$ to turn it into a DE, and then used our R-K method to solve it, it would give an exact answer even for arbitrarily large h. What this means (shades of Taylor series) is that this method can follow quite a bit of curvature accurately. (If you actually *had* that particular DE, you could solve it analytically. However, that analysis shows that our R-K method is like using the first four terms of a Taylor series for the derivative.)

2. For what kind of environmental problems would Runge-Kutta methods be used?

   • The method is used for solving systems of (and sometimes single) differential equations that model:

   a) transport and fate of materials moving around among multiple compartments. Two examples are pollution transport in streams, and phosphorus exchanges among water, sediments, and various species (or trophic levels) in lakes.

   b) models for interacting populations in fisheries and wildlife biology, including risk analyses for endangered species.

c) chemical reactions among multiple, interacting chemicals.

d) the various "reacting vessels" in a sewage-treatment plant.

There are many more. I am surprised by how often people seem to use Euler's method for solving such problems, given that Runge-Kutta is *so* much more accurate.

3. "Why have we been applying numerical methods to differential equations that can be solved analytically?"

• Only to test and illustrate how the methods work with equations for which we know the true solution—we use that true solution for comparison. In practice, the analytic solution would always be preferable, since it would give exact, and general, results. However, there are lots of DEs (and especially systems of DEs) for which analytic solutions are not available, so Runge-Kutta is used a lot.

4. How do you recognize the need to solve ODEs numerically as opposed to analytically? A lot of them look too hard to solve analytically, but should one try first (possibly using software), and only turn to R-K and Euler if that doesn't work?

• Please, never use Euler, since you can use R-K! I suggest that you first look in tables or try MATLAB, or both. If you can't get an analytic solution by one of those methods, then turn to Runge-Kutta.

5. In Runge-Kutta, are there only four calculations for slope, never more?

• With the particular R-K method we work with, you calculate four slopes, and their weighted average for each step and for each equation in the system. Quoth the raven....

6. Can our 4th-order R-K method be extrapolated to higher orders, for more accurate solutions? If so, would the slopes be weighted as in Simpson's rule, i.e., 1 4 2 4 2 4 ... 4 1?

• There are higher order versions (which we won't study). In fact, the variable-time-step method used in MATLAB's "ode45" turns out to be 5th order. I think the weightings would not then be those of Simpson's rule, which is third order, and does a different thing.

7. What kinds of problems would you use the R-K method for?

• I have described two cases; i.e., the modified (and more realistic?) Streeter-Phelps equations, and following flows of water (and bacteria, and various substances) through reservoirs. DEs describing reactions among numerous chemical substances (like those producing the "ozone hole" in the atmosphere) would be another situation, if you wanted to deal with non-equilibrium cases. The "SIR" model for epidemics is another example.

8. If an equation can't be solved analytically, how can it be solved by Runge-Kutta?

• An analytic solution gives a very general formula for the *exact* solution. That is very desirable, but if you can't get that, then R-K just gives you a list of approximate y(t=0), y(t=h), y(t=2h), ... points. It does that by the process of going to various points that are approximations to the true values, and using the slopes at those "guesswork" points to estimate the true slope. So R-K doesn't yield a true solution—it just gives a series of approximations.

The situation here is closely related to the fact that we noted earlier, that some integrals yield analytic results but others do not.

9. Why in Runge-Kutta do we go in the sequence $h/2, h/2, h$?

• The idea is to use the point at the left (starting) end of the step once, a point at the right (ending) end of the step once, and points in the middle of the range twice. It's a bit like the 1-4-1 weighting of Simpson's rule, in that points near the center of the range are treated as more representative than those at the ends.

Chapter 8

Second-Order ODEs

Because they are important in their own right, and because many important *partial* differential equations (Chapter 11) are second order, we now take up second-order *ordinary* differential equations. This means that we must deal with second derivatives, sometimes in combination with first derivatives, and sometimes alone. We will be concerned with equations of the general form

$$f_1(x,y)\frac{d^2y}{dx^2} + f_2(x,y)\frac{dy}{dx} + f_3(x,y) = 0. \tag{8.1}$$

Many of the equations that arise in applications take the simpler form of the *linear* second-order equation[1],

$$f_1(x)\frac{d^2y}{dx^2} + f_2(x)\frac{dy}{dx} + f_3(x)y + f_4(x) = 0. \tag{8.2}$$

In environmental science, this form of mathematical model arises most often in describing various transfers of energy (as heat) and mass. Mass transfers of interest might include SO_2 and NO_x as air pollutants, CO_2 to a forest canopy, water from a leaf or from a body of water, and groundwater flows. As we shall see, one fundamental principle that applies for most of these tranfers is that exchange rates are proportional to concentration, temperature, or pressure *differences*; that is to *gradients* (spatial derivatives) of those quantities.

---

[1]Can you explain to a colleague just why Eqn. 8.2 is conceptually much simpler than Eqn. 8.1?

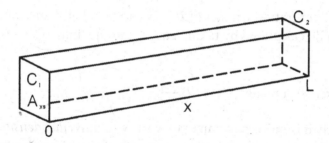

Figure 8.1: Diagram of a hollow tube used to derive the diffusion equation.

8.1 Mass Transfer in Cartesian Coordinate Systems

We begin with an example of mass transfer—diffusion of a gas down a long rectangular tube, as shown in Fig. 8.1. A gas will diffuse whenever there is a concentration difference[2] (i.e., when $C_1 \neq C_2$) from one end of the tube to the other. (We are assuming here the simple case where a single gas, such as chloroform, is diffusing at relatively low concentrations through air, without participating in any reactions and without being adsorbed onto the walls of the tube along the way.) Under these conditions, R, the mass flow rate (or *flux*) in the $+x$ direction, is given in units like mol s^{-1} by

$$R = D\left(\frac{C_1 - C_2}{L}\right) A_{xs} = -A_{xs}D\left(\frac{C_2 - C_1}{L}\right), \tag{8.3}$$

where C_1 and C_2 are chloroform concentrations [mol cm^{-3}] at the two ends of the tube, L is the tube length [cm], A_{xs} is its cross-sectional area [cm^2], and D is the molecular diffusivity of chloroform through air [cm^2 s^{-1}].
Thus, Eqn 8.3 has consistent units of

$$\left[\frac{mg}{s}\right] = \left[\frac{cm^2}{s}\right]\left[\frac{mg}{cm^3 cm}\right] cm^2.$$

Eqn. 8.3 gives the mass flow rate down a tube of finite length L. Often we will need to know what R would be across a plane, in response

---

[2]Strictly speaking, gases diffuse in response to differences in mass fractions (Parkhurst 1994), but these are proportional to concentrations in isobaric (uniform pressure) systems.

to the concentration *gradient* (dC/dx) across that plane. Such flow rates can be calculated by taking Eqn. 8.3 in the limit as $L \to 0$, which yields

$$R \text{ across a plane } = -A_{xs}D\frac{dC}{dx}. \qquad (8.4)$$

Frequently it is more convenient to work with *mass flux densities*, with dimensions of mass area^{-1} time^{-1}, rather than with mass fluxes or flow rates [mass time^{-1}]. For our present problem, the flux density, in mg cm^{-2} s^{-1}, of gas diffusing across the plane is given by

$$F = R/A_{xs} = -D\frac{dC}{dx}, \qquad (8.5)$$

which is known as *Fick's first law of diffusion*, and dates back to at least 1855 (Nobel 1991). The minus sign on the RHS of Eqn. 8.5 is puzzling to many at first, but it must be present because (1) diffusion moves material from high concentration regions to low concentration regions (i.e., down a gradient), and (2) defining the flux to be positive when material moves in the positive x direction is a useful convention that we adopt.

A unit check for Eqn. 8.5 takes the form

$$\frac{mg}{cm^2\, s} = \frac{(mg/s)}{cm^2} = \frac{cm^2}{s}\frac{mg}{cm^3}\frac{1}{cm}.$$

Next we take up an example where we must work with this derivative form. Suppose we consider a specific short section of our rectangular tube, which we will refer to as a "control volume." Also suppose that some gas is lost from the air in the tube as the gas diffuses along, by adsorption onto (adhering to) the tube wall. We might model this adsorption using

$$W = kPC \qquad (8.6)$$

where W is the adsorption rate of the gas per unit length of tube [mg cm^{-1} s^{-1}], $P = 4S$ is the perimeter of the tube [cm] (comparable to the circumference of a circular cylinder), k is a rate coefficient [cm s^{-1}] known as a *deposition velocity* because of its velocity-like units, and C is the chloroform concentration at x. I leave the reader to perform a unit check for Eqn. 8.6.

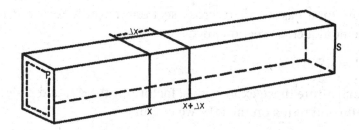

Figure 8.2: Diffusion tube showing "control volume," of length Δx. The perimeter P is the sum of the lengths of the four sides indicated.

For a steady state situation, we work with a mass balance over any convenient period of time for the control volume (CV) shown in Fig 8.2. This takes the form

(diffusion rate in at x) = (uptake rate by wall within CV)

+ (diffusion rate out at $x + \Delta x$).   (8.7)

It is not obvious how C will vary with x along the tube in this scenario. (Can you guess?) But we can work it out. Similarly to the situations in Chapter 4, we have a case in which our fundamental knowledge does not tell us $C(x)$ directly, but rather has to do with rates of change along x. A straightforward mass balance (assuming a steady state in time) will lead us to the appropriate differential equation, whose *solution* (as before) will give us $C(x)$.

Eqn. 8.7 represents a balance for the movement of chloroform mass into and out of the control volume over some period of time Δt. This states, if $C(x)$ is not changing with time, that mass m diffusing across the plane at x is balanced by the sum of mass diffusing out at $x + \Delta x$ plus mass adsorbing onto the tube wall:

$$-DA_{xs}\frac{dC}{dx}\bigg)_x \Delta t \approx -DA_{xs}\frac{dC}{dx}\bigg)_{x+\Delta x} \Delta t + kPC\Delta x \Delta t. \qquad (8.8)$$

A unit check yields

$$\left[\frac{cm^2}{s}\right]cm^2\left[\frac{mg}{cm^4}\right]s = \left[\frac{cm^2}{s}\right]cm^2\left[\frac{mg}{cm^4}\right]s$$

$$+ \left[\frac{cm}{s}\right]cm\left[\frac{mg}{cm^3}\right]cm\ s.$$

If the tube has a square cross section of width S, and P is the perimeter of a cross section, then

$$P = 4S \text{ and } A_{xs} = S^2.$$

If we substitute these expressions, factor out the Δt, and collect the two diffusion terms on the LHS, we obtain

$$DS^2 \left[\left. \frac{dC}{dx} \right)_{x+\Delta x} - \left. \frac{dC}{dx} \right)_x \right] \approx 4kSC\Delta x$$

or

$$\frac{\left. \frac{dC}{dx} \right)_{x+\Delta x} - \left. \frac{dC}{dx} \right)_x}{\Delta x} \approx \frac{4kC}{DS}. \tag{8.9}$$

Next, to help keep things straight, it can be helpful to make a temporary substitution[3], using G to represent the concentration *gradient*:

$$G \stackrel{\text{def}}{=} \frac{dC}{dx}.$$

Then Eqn. 8.9 becomes

$$\frac{G_{x+\Delta x} - G_x}{\Delta x} \approx \frac{4kC}{DS}.$$

As we have seen in previous ODE derivations, this expression is only approximate because C (and probably G) varies somewhat along the Δx length of the control volume. So, we take the limit as $\Delta x \to 0$ to obtain

$$\lim_{\Delta x \to 0} \frac{G_{x+\Delta x} - G_x}{\Delta x} = \frac{dG}{dx} = \frac{4kC}{DS}.$$

But from the definitions of G and of second derivatives,

$$\frac{dG}{dx} = \frac{d}{dx} \left(\frac{dC}{dx} \right) = \frac{d^2C}{dx^2}.$$

Finally, then, we obtain[4]

---

[3]This type of substitution is not *necessary*, and once you've had a little practice with second-order equations, you'll probably go directly to the result without it.

[4]Using a^2 as the constant turns out to be convenient for two reasons—first, it indicates that the constant is positive. Second, the solution to Eqn. 8.10 will turn out to contain the square root of this constant, so if we make it a square, the solution is simplified.

$$\frac{d^2C}{dx^2} = \frac{4kC}{DS} = a^2C \text{ if } a^2 \overset{\text{def}}{=} \frac{4k}{DS}. \tag{8.10}$$

A unit check here yields

$$\left[\frac{mg}{cm^3}\right]\left[\frac{1}{cm^2}\right] = \left[\frac{cm}{s}\right]\left[\frac{mg}{cm^3}\right]\left[\frac{s}{cm^2}\right]\left[\frac{1}{cm}\right].$$

This is a second-order ODE, one that represents our knowledge of the balance of chloroform mass transfers within the tube. To find $C(x)$, we would want to solve this equation.

Solving second-order ODEs involves integrating twice, so two constants of integration arise; thus, two equivalents of initial conditions are needed. For problems like this one, where time is not a variable, the two fixed conditions might be values of C_1 and C_2, for example, and these are called *boundary conditions* (BCs). Here $C(0) = C_1$ and $C(L) = C_2$.

Murphy (1960) has a chapter covering solutions of second-order equations; in it, $y'' = a^2y$ is the one that matches our diffusion equation. Murphy presents two alternate solutions that in our context take the form,

$$C(x) = A\sinh ax + B\cosh ax, \text{ or} \tag{8.11}$$

$$C(x) = Ae^{ax} + Be^{-ax}, \tag{8.12}$$

and we may choose whichever form provides the more convenient solution. In either case, one substitutes the two boundary conditions in turn into the chosen solution, yielding two expressions that can be solved for the constants A and B.

Suppose that for some reason we needed to solve our differential equation numerically. This need might arise if the dimensions of the tube varied with x, for example. Then we could translate our one second-order equation into a system of two first-order ones using

$$\frac{dC}{dx} \overset{\text{def}}{=} G, \text{ (the gradient), and}$$

$$\frac{dG}{dx} = \frac{4kC}{DS}.$$

This system could be solved using a variant of our Runge-Kutta scheme called a *shooting method* (Press et al. 1992) to deal with the fact that we don't know both G and C at $x = 0$. Anyway, any second-order ODE can be converted into a pair of first-order equations in a similar manner.

Some Example Analytic Solutions

For a long square tube with a gas diffusing along its length and adsorbing onto its walls (at a rate proportional to the local concentration), we found that

$$\frac{d^2C}{dx^2} = \frac{4kC}{DS} = a^2C.$$

We also found the general solution to this equation (in two forms) from Murphy's tables. Now let's look at some particular solutions of this equation for a few special cases:

1. First, suppose that gas does *not* adsorb onto the tube wall, so that $k = 0$. Also suppose that C_1 and C_2 are two known constants. Here

$$\frac{d^2C}{dx^2} = \frac{d}{dx}\left(\frac{dC}{dx}\right) = 0. \qquad (8.13)$$

To solve this form, we can treat x as one variable and dC/dx as another, separate the two variables, and integrate once to obtain[5]

$$\int d\left(\frac{dC}{dx}\right) = \int 0 \, dx \Rightarrow \frac{dC}{dx} = a,$$

where a is a constant of integration. This latter expression can be solved by a new separation of variables and another integration:

$$\frac{dC}{dx} = a \Rightarrow \int dC = \int a \, dx \Rightarrow C = ax + b, \qquad (8.14)$$

where b is a second constant of integration. This is, of course, consistent with a linear function (Eqn. 8.14) having a zero second derivative (Eqn. 8.13).

Note the simple result here—given diffusion down the tube with different, fixed concentrations at the two ends and with no sources or sinks, the gas concentration varies linearly along the length. To solve for the constants of integration, we note that $C = ax + b$ holds at $x = 0$ and at $x = L$, from which

$$C(0) = C_1 = a0 + b \Rightarrow b = C_1$$

---

[5]The symbol "$\Rightarrow$" is read "implies." That means, in turn, "It must follow that."

$$C(L) = C_2 = aL + b \Rightarrow a = \frac{C_2 - C_1}{L}.$$

Thus the complete solution becomes $C(x) = C_1 + (C_2 - C_1)(x/L)$.

As always, it is good practice to be sure our solution satisfies both the DE and the BCs. The latter checks are easy, since $C(0) = C_1$ and $C(L) = C_2$ by simple substitution of $x = 0$ and $x = L$ into the solution. To check that we have satisfied $C'' = 0$, we simply differentiate the solution twice. Clearly $C' = a$, from which $C'' = 0$, so the check is complete.

2. Next we solve the problem as originally stated, with $k \neq 0$, $C(0) = C_1$, and $C(L) = C_2$. If we set[6] $a^2 \stackrel{\text{def}}{=} 4k/(DS)$, our DE becomes

$$C'' = \frac{4k}{DS} C = a^2 C.$$

As before, using a^2 rather than a provides a more convenient solution, and it also represents a positive constant; this is consistent with $4k/DS$ being positive.

I stated above (p. 171) that both $C(x) = A\exp(ax) + B\exp(-ax)$ and $C(x) = A\sinh ax + B\cosh ax$ are solutions to our DE, and Exercise 1, p. 183, asks you to prove that these expressions do satisfy $C'' = a^2 C$. If we choose the second form, and use the BCs to solve for A and B, we obtain

$$C(0) = A\sinh 0 + B\cosh 0 = C_1$$

$$C(L) = A\sinh aL + B\cosh aL = C_2.$$

Because[7] $\sinh 0 = 0$ and $\cosh 0 = 1$, the first of these equations yields $B = C_1$. Simple algebra applied to the second then yields

$$A = \frac{C_2 - C_1 \cosh aL}{\sinh aL}.$$

The overall solution becomes

$$C(x) = (C_2 - C_1 \cosh aL)\left(\frac{\sinh ax}{\sinh aL}\right) + C_1 \cosh ax, \tag{8.15}$$

which I leave for you to verify. For example, is it correct at $x = 0$ and at $x = L$? One such solution is shown in Fig. 8.3 (left), p. 174.

---

[6]This form for a^2 applies to a tube with a square cross section; the coefficient would be different for other shapes.

[7]You could check these values using tables, a calculator, or the definitions of the sinh and cosh functions.

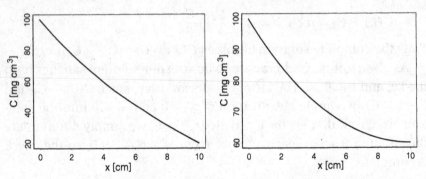

Figure 8.3: Plots of Eqn. 8.15 when $C_1 = 100$ and $C_2 = 50$ (left), and of Eqn. 8.16 with $C_1 = 100$ and $dC/dx = 0$ at $x = 10$ (right). For both, $a = 0.11$. Note that in the first case, the slope of the relationship does not go to zero at $x = L$, while in the second case it does go to zero.

3. Again suppose $k \neq 0$ and $C(0) = C_1$, but now let us specify that the tube has an impermeable cap at $x = L$; i.e., that no chloroform diffuses across the plane there. This latter BC requires that

$$-D\frac{dC}{dx} = 0 \text{ at } x = L, \text{ from which } \left.\frac{dC}{dx}\right)_L = 0.$$

Our BCs are now

$$C(0) = A\sinh 0 + B\cosh 0 = A(0) + B(1) = C_1 \text{ and}$$

$$\left.\frac{dC}{dx}\right)_L = \frac{d}{dx}(A\sinh ax + B\cosh ax)_L =$$

$$aA\cosh aL + aB\sinh aL = 0.$$

The first BC implies as before that $B = C_1$, while the second reduces to

$$A = -C_1\frac{\sinh aL}{\cosh aL} = -C_1\tanh aL,$$

where $\tanh x \stackrel{\text{def}}{=} \sinh x / \cosh x$ (by definition). The final solution is thus

$$C(x) = C_1(\cosh ax - \sinh ax \tanh aL). \tag{8.16}$$

I leave this solution for you to check, but close by plotting its general form in Fig. 8.3 (right). Note from the plot that the curve becomes level at $x = L$, consistent with $dC/dx = 0$ there.

We have now laid the groundwork for deriving some important *partial* differential equations, which will be the subject of Chapter 11.

8.2   Generalizations to Other Quantities

Concepts like those represented by the equation

$$R = -DA\frac{dC}{dx}, \tag{8.17}$$

which indicates that transfers of mass by diffusion across a plane occur at rates proportional to the concentration gradient and proportional to the area available, are very general ones in the physical world. The same general concept applies to transfers of heat by conduction, of groundwater, of electrical charge, and of organisms modelled as moving by Brownian-like random processes. Some of the chapter exercises deal with the latter situation (sometimes across a line rather than across a plane).

The analogous relationships to Eqn. 8.17 for some of these processes are given here. For the *discharge Q* of groundwater [e.g., in $m^3\ hr^{-1}$],

$$Q = -KA\frac{dh}{dx}, \tag{8.18}$$

where h is the *hydraulic head* (a pressure, often expressed in m of water) and K is the *hydraulic conductivity* [e.g., in m hr^{-1}]. This relationship is known as *Darcy's law*. The coefficient K depends on the properties of the flowing fluid and of the matrix, such as soil, through which it flows.

For organisms moving on a planar surface, and modelled as diffusing across a line (rather than across a plane), the relationship might take the form

$$M = -DL\frac{dC}{dx}, \tag{8.19}$$

with M being the migration rate [organisms da^{-1}], L being the length of the line across which the organisms can move, C being the organism density [number per m^2], and D is the "diffusivity" (or "migrativity," if you will) of the organisms on that surface [$m^2\ da^{-1}$]. A similar situation is the basis for Exercise 2, p. 183, except that it deals with fish moving through 3-dimensional space (they aren't confined to a surface), and they move across a plane like our diffusing gas molecules, rather than just across a line.

For conduction of heat through a solid, a standard model is

$$Q = -kA\frac{dT}{dx}, \tag{8.20}$$

with Q the heat flux [J s^{-1}], T the temperature [deg C], and k the thermal conductivity [J deg^{-1} m^{-1} s^{-1}].

You may find it useful to check the units for each of Eqns. 8.18–8.20.

8.3 Conduction of Heat in a Solid

Now we consider just one of those examples, to serve as a variation on the theme of diffusion that we have already dealt with. In particular, we will derive the second-order ODE that describes conduction heat transfer along the length of a small cylinder. We'll take the case of roasting hot dogs or marshmallows, using a heavy steel wire as our "spit."

Suppose we have a long, thin cylinder of radius r, and we hold one end in a fire so that $T(0) = T_1$, and hold the other end in our hand, where $T(L) = T_2$. (We hope that T_2 doesn't go much above 35 C!) Let us suppose that heat not only moves along the wire by conduction, but is also lost to the air by the process of convection. Specifically, a good model for that process would be that from any short section of the wire where the temperature is $T(x)$ deg, heat loss to the air proceeds at a rate Q_{surf} [J s^{-1}] given by

$$Q_{surf} = hA_s(T(x) - T_a), \tag{8.21}$$

with h the *convection coefficient* [J cm^{-2} deg^{-1} s^{-1}], A_s the surface area of the wire in contact with the air [cm^2], and T_a the air temperature [deg]. (We will ignore the complication that in reality, the air temperature would be higher near the fire than it is near your hand, and treat it as a constant here.)

The law of conservation of energy then tells us that in a steady state, for any small length of the wire between x and $x + \Delta x$,

Energy conducted in at x = Energy conducted out at $(x + \Delta x)$

+Energy lost to air.

We use Eqn. 8.20 for the first two terms and Eqn. 8.21 for the third to obtain

$$-kA_{xs}\frac{dT}{dx}\bigg)_x \approx -kA_{xs}\frac{dT}{dx}\bigg)_{x+\Delta x} + hA_s(T(x) - T_a), \qquad (8.22)$$

with $A_{xs} = \pi r^2$ being the cross-sectional area of the rod (across which conduction occurs) and $A_s = 2\pi r\Delta x$ being the surface area of the Δx-long section that is exposed to the air. Can you explain why this equation is only approximate?

Before we go further, it's time for a unit check:

$$\frac{J}{cm\ s\ deg}cm^2\frac{deg}{cm} \approx \frac{J}{cm\ s\ deg}cm^2\frac{deg}{cm} + \frac{J}{cm^2\ s\ deg}cm^2\ deg.$$

Does this check?

We now manipulate Eqn. 8.22 by operations similar to those we used to move from Eqn. 8.10 to Eqn. 8.19. We have

$$kA_{xs}\left[\frac{dT}{dx}\bigg)_{x+\Delta x} - \frac{dT}{dx}\bigg)_x\right] \approx hA_s(T(x) - T_a),$$

$$\left[\frac{dT}{dx}\bigg)_{x+\Delta x} - \frac{dT}{dx}\bigg)_x\right] \approx \frac{hA_s}{kA_{xs}}(T(x) - T_a),$$

$$\left[\frac{dT}{dx}\bigg)_{x+\Delta x} - \frac{dT}{dx}\bigg)_x\right] \approx \frac{h2\pi r\Delta x}{k\pi r^2}(T(x) - T_a),$$

or

$$\frac{\left[\dfrac{dT}{dx}\bigg)_{x+\Delta x} - \dfrac{dT}{dx}\bigg)_x\right]}{\Delta x} \approx \frac{2h}{rk}(T(x) - T_a).$$

Finally, we take the limit as $\Delta x \to 0$.

$$\lim_{\Delta x \to 0}\frac{\left[\dfrac{dT}{dx}\bigg)_{x+\Delta x} - \dfrac{dT}{dx}\bigg)_x\right]}{\Delta x} = \frac{2h}{rk}(T(x) - T_a).$$

If you find it helpful, you could define $G = dT/dx$—the temperature gradient—and work with that, but in any case the result will be

$$\frac{d^2T}{dx^2} = \frac{2h}{rk}(T - T_a).$$

The easiest way to *solve* that equation (if T_a is a constant) is to define $\theta \overset{\text{def}}{=} T - T_a$, and to note that

$$\frac{d\theta}{dx} = \frac{d(T - T_a)}{dx} = \frac{dT}{dx} - \frac{dT_a}{dx} = \frac{dT}{dx},$$

because the derivative of T_a would be zero. The same argument holds for the second derivative, and so

$$\frac{d^2\theta}{dx^2} = \frac{2h}{rk}\theta.$$

Now if we set $a^2 = 2h/(rk)$, we have the same equation we already solved in the diffusion context, but with temperature difference θ replacing concentration as the *field variable*.

8.4 Coordinate Systems for Curved Geometry

Spherical Geometry

Reminder: mass flux density equals $-D(dC/dx)$, with dimensions of mass time^{-1} area^{-1}. In a similar way, heat flux density equals

$$q \;[\text{J s}^{-1}\text{m}^{-2}] = -k\frac{dT}{dx} = \left[\frac{\text{J}}{\text{deg m s}}\right]\left[\frac{\text{deg}}{\text{m}}\right],$$

where k is the *thermal conductivity* of a substance, and T is temperature.

If we are interested in conductive heat transfer in part of a spherical object, e.g., in the fur of a small "rolled up" animal, it is most convenient to work in spherical coordinates rather than in Cartesian (x-y-z) coordinates. Spherical coordinates are often defined as shown in Fig. 8.4, p. 179.

We will consider a simplified situation in which there is no angular variation of temperature or other features of the sphere, so that temperature and heat flux densities vary only with radius r. We will derive the energy-balance equation for this situation based on the Δr control volume, a thin, spherical shell in the fur, diagrammed as in Fig. 8.5, p. 179. (The thickness of the fur, relative to the body diameter, is exaggerated in that diagram.)

For this process, it will be useful to remember that the surface area A and the volume V of a sphere of radius r are given by $A = 4\pi r^2$

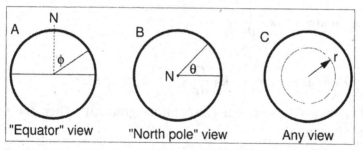

Figure 8.4: Spherical coordinates are most easily seen in an analogy to the earth. The angle ϕ corresponds roughly (though not exactly) to latitude, while θ corresponds roughly with longitude. Any point within, or on, the sphere can be specified uniquely by its ϕ, θ, and r coordinates. The N symbol indicates the point that would be the "north pole" if this sphere were our globe.

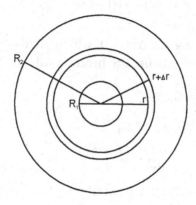

Figure 8.5: Cross section through a sphere showing a control volume bounded by two concentric spheres.

and $V = 4\pi r^3/3$. It may also help to notice that the surface area is the r-ward derivative of the volume ($dV/dr = 4\pi r^2 = A$) and that $dV = A\,dr$.

In a steady-state situation[8], when no heat is produced in the fur, the heat entering any "shell" of fur must equal the heat leaving that shell in any given period of time. Thus, the *heat flux Q* in at r equals Q out at $r + \Delta r$, where

$$Q \left[\frac{J}{s}\right] = qA \left[\frac{J}{s\,m^2}\right] [m^2],$$

---

[8]I.e., when nothing changes with time

$$Q \text{ in at } r = -k \left(\frac{dT}{dr}\right)_r A(r),$$

and

$$Q \text{ out at } r + \Delta r = -k \left(\frac{dT}{dr}\right)_{r+\Delta r} A(r + \Delta r).$$

Let $dT/dr \overset{\text{def}}{=} G$, the radial temperature gradient. Then the energy balance becomes

$$-kG(r)A(r) = -kG(r + \Delta r) A(r + \Delta r), \text{ or}$$

$$k\left[G(r + \Delta r) A(r + \Delta r) - G(r)A(r)\right] = 0.$$

Since k is not zero, we can divide through by $k\Delta r$ to simplify the equation to

$$\frac{G(r + \Delta r) A(r + \Delta r) - G(r)A(r)}{\Delta r} = 0.$$

Then, we take the limit as $\Delta r \to 0$, which yields $d(GA)/dr = 0$. Expanding that derivative using the product rule (p. 25) leads to

$$A\frac{dG}{dr} + G\frac{dA}{dr} = 0,$$

or, if we substitute for A and G,

$$4\pi r^2 \frac{d}{dr}\left(\frac{dT}{dr}\right) + \frac{dT}{dr} \cdot \frac{d\left(4\pi r^2\right)}{dr} = 0.$$

Next we divide through by 4π to obtain

$$r^2 \frac{d^2T}{dr^2} + \frac{dT}{dr} \cdot \frac{dr^2}{dr} = 0, \text{ or}$$

$$r^2 \frac{d^2T}{dr^2} + 2r\frac{dT}{dr} = 0, \text{ or finally}$$

$$\frac{d^2T}{dr^2} + \frac{2}{r}\frac{dT}{dr} = 0.$$

The solution to this equation can be found in Murphy (1960) or in Carslaw and Jaeger (1947), or by using software like MATLAB®, Maple®, Mathematica®, or Octave. For example, if $T(R_1) = T_1$ and $T(R_2) = T_2$, the solution is

$$T(r) = \frac{1}{R_2 - R_1}\left[(T_2 R_2 - T_1 R_1) + \frac{R_1 R_2}{r}(T_1 - T_2)\right], \quad (8.23)$$

which is of the form $T(r) = a + b/r$, and is plotted in Fig. 8.6 (left), p. 181, for the boundary conditions indicated.

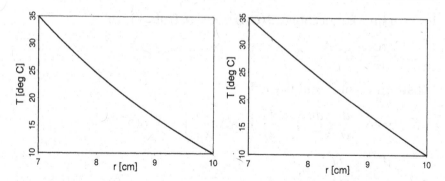

Figure 8.6: Temperature profiles in the fur of a spherical animal (left) and a cylindrical animal (right), with body diameters of 7 cm and fur thicknesses of 3 cm more. In each case, the body temperature is 35° C, and the temperature at the outer fur surface is 10° C. The result for the sphere is given by Eqn 8.23, p. 180, and that for the cylinder by Eqn 8.24, p. 182.

Cylindrical Geometry

Fig. 8.7 defines the components of the cylindrical coordinate system. In the steady state, Q in at $r = Q$ out at $r + \Delta r$. The equation for that energy balance is

$$-k \left. \frac{dT}{dr} \right)_r A(r) = -k \left. \frac{dT}{dr} \right)_{r+\Delta r} A(r + \Delta r)$$

Here, as with the sphere, we will consider temperature variation only in the radial dimension—refer to Fig. 8.5, (p. 179), which can also be interpreted as showing the end of a cylinder. Again, let $dT/dr \overset{\text{def}}{=} G$. Then, as with the sphere,

$$k \left[G(r + \Delta r) A(r + \Delta r) - G(r)A(r) \right] = 0.$$

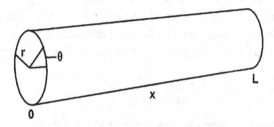

Figure 8.7: Components of the cylindrical coordinate system.

In the limit, that becomes $d(GA)/dr = 0$, from which

$$A\frac{dG}{dr} + G\frac{dA}{dr} = 0.$$

Substituting for A and G yields

$$2\pi r \Delta z \frac{d}{dr}\left(\frac{dT}{dr}\right) + \frac{dT}{dr}\left[2\pi \Delta z \frac{dr}{dr}\right] = 0, \text{ or}$$

$$r\frac{d^2 T}{dr^2} + \frac{dT}{dr} = 0,$$

with boundary conditions $T(R_1) = T_1$ and $T(R_2) = T_2$.

The solution to this equation (again from tables) is

$$T(r) = \frac{1}{\log(R_2/R_1)}\left[(T_1 \log R_2 - T_2 \log R_1) - (T_1 - T_2)\log r\right], \quad (8.24)$$

which is of the form $T(r) = a + b \log r$. The appearance of $\log r$ in this equation precludes specifying a fixed-temperature boundary condition at $r = 0$, because the log of zero is undefined. The solution is shown in Fig. 8.6 (right), p. 181, for the boundary conditions indicated. Note the similarity to the solution for the sphere—the general shape is the same, but the slope and curvature differ with geometry.

We will see these coordinate systems again in Chapter 11, where we will allow concentrations and temperatures to vary with time as well as with radius.

8.5 Solving Second-Order ODEs Analytically with MATLAB

When a second-order ODE has an analytic solution, MATLAB can probably find it. For example, to solve Eqn. 8.10 with $C(0) = C_1$ and $C'(L) = 0$, as with Case 3 on p. 174, we could enter
`C=dsolve('D2Conc=a^2*Conc','Conc(0)=C1','DConc(L)=0'),`
where the D2Conc denotes the second derivative of Conc. Then we could enter `pretty(simple(C))`[9] to see the answer in the form

$$C_1 \frac{\exp[a(x - L)] + \exp[-a(x - L)]}{\exp(aL) + \exp(-aL)}.$$

---

[9]MATLAB's "simple" function tries various ways to simplify an expression, and returns the shortest one.

A bit of algebra shows that to be equal to the Eqn. 8.16 that we obtained earlier (p. 174). As with other commands, you can obtain more information about using MATLAB to solve this kind of equation by entering help dsolve.

8.6 Exercises

1. Differentiate each of the putative solutions in Eqns. 8.11 and 8.12, (p. 171) twice, to show that both satisfy $C'' = a^2 C$.

2. Ecologists sometimes model animal migration by analogy with diffusion, and this exercise follows that approach. Consider a stream in which fish are kept stocked at a (roughly) constant density of F_0 fish per cubic meter of water at one position $x = 0$ along a stream. In the next 1000 m downstream (i.e., for $0 \leq x \leq 1000$), fishing is not allowed. Then, at $x = 1000$ m, fishing pressure keeps the fish density at a lower, but nearly constant value F_1. In the "no-fishing" reach of the stream, there is a little natural reproduction, but deaths exceed births there so that in any small volume of water in this reach, the instantaneous net rate of loss of fish (deaths minus births) is $L\%$ per year.

   At any (imaginary) vertical plane across the stream, the fish migrate across the plane in a diffusion-like fashion, at a rate equal to

$$M = -A_x D \frac{dF}{dx},$$

   where M is the migration rate or flux [fish per day] in the positive (downstream) x direction, A_x is the (assumed uniform) cross-sectional area of the stream [m^2], D is the "diffusivity" [m^2 da^{-1}], and t is time [da]. The diagram of the square tube on p. 167 may illustrate this situation as well, if you replace C with F.

   Assuming that the fish density remains constant through time at any distance x along the unfished reach, derive the differential equation that, if solved, would tell us the fish density at any point along that reach. State the boundary conditions clearly, as well. Solve this equation for $F(x)$, and determine what the density is in the center of the reach, i.e., at $x = 500$ m.

3. Temperatures at various depths in a soil profile can affect organisms living there, the rates of decomposition of organic matter there, the emission of methane as a result of such decomposition, and many other processes. Such temperatures nearly always vary with both depth and with time, and *partial* differential equations (PDEs) are then necessary to model the variations. Here we model the somewhat artificial process of steady-state (i.e., time-independent) conduction of heat in soil, to set the stage for modelling more realistic situations when we study PDEs.

   Conduction of heat in a solid occurs by a process analogous to molecular diffusion, with (as I understand the process) molecular vibrations and electrons zipping around somewhat randomly, with more activity in higher temperature regions than in cooler ones. Temperature can be thought of, roughly anyway, as a measure of energy concentration. Thus, heat energy is conducted from high to low temperature regions, and moves at rates proportional to temperature gradients. In a close analogy to molecular diffusion of some mass through a fluid, the conductive heat flux Q across a plane is given by

$$Q = -kA\frac{dT}{dx} \quad [\text{J s}^{-1}],$$

   and the heat flux density q is given by

$$q = -k\frac{dT}{dx} \quad [\text{J m}^{-2}\,\text{s}^{-1}].$$

   Here k is the *thermal conductivity* $[\text{J m}^{-1}\,\text{s}^{-1}\,\text{deg}^{-1}]$, A is the area of the plane across which the flux passes, and x is the direction normal (perpendicular) to that plane.

   Suppose the surface temperature of the soil (perhaps that under a building) stays constant at a temperature T_0, and the temperature at a depth of L meters stays constant at a temperature T_L. (As noted above, this is a somewhat artificial situation.) Also suppose that microbes in this soil produce heat (from their metabolic processes) at a uniform rate of M $\text{J m}^{-3}\,\text{s}^{-1}$. Derive the differential equation whose solution would give the temperature at any depth between depths $z = 0$ and $z = L$. Then solve the equation for $T(z)$. You'll probably find it helpful to think of the column of soil under a representative square meter of surface.

4. Derive the second-order differential equation for gas diffusion through a *circular* tube with gas adsorption at the tube wall. The tube has length L and radius R. The gas diffusivity through air is D cm^2 s^{-1} and the adsorption rate coefficient is k cm s^{-1}. (Note: Those units can be thought of as mg adsorbed per second per mg cm^{-3} of concentration). The gas concentrations on the ends of the tube are C_1 and C_2 mg cm^{-3}, respectively. This problem is similar to the one discussed at the start of the chapter on second-order ODEs, except that the tube here has a circular cross-section instead of a rectangular one. (Assume that the concentration varies only along the length, but not with radius. Thus a Cartesian coordinate system is adequate for this problem.)

5. Consider two small sub-arctic islands, A and B, connected by an isthmus L m long and W m wide. (Fig. 6.1, p. 142, can be used here.) There are voles on each, with constant population densities ρ_A and ρ_B [voles m^{-2}], respectively. (In reality, the densities would likely vary with time. We can deal with that complication when we get to Chapter 11.) Those constants form the boundary conditions for the density variation across the isthmus, and you may take them as given. The voles migrate along the isthmus in a diffusion-like process, with the migration rate M voles da^{-1} across any line of width w being

$$M = -Dw\frac{d\rho}{dx}.$$

The "migrativity" D is in m^2 da^{-1}, and x [m] is the distance from Island A along the isthmus. Voles are exposed as they move along the isthmus, and a fraction f of those on any small area are taken per day by predatory birds. (Treat that as an instantaneous rate.)

Derive the differential equation that if solved would tell us how the vole density varies along the length of the isthmus.

6. When you study systems of linear algebraic equations, you will work with heat loss through a layer of insulation on a domestic water heater. Those calculations will be simplified by neglecting effects of curvature of the walls of that water heater. This problem takes the curvature into account.

Consider a hot water heater with a boiler of diameter D. It is constructed of steel with a thin glass lining. The steel is then surrounded by a thickness I of insulation, and finally by a second layer of steel. Fig. 8.8 helps to define some of the variables referred to below.

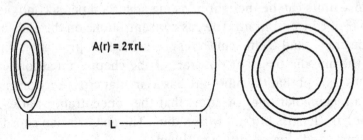

Figure 8.8: Side view of the water heater (tipped on its side) (left), and end view (right) with the thickness of insulation exaggerated. The area formula shown applies to the curved side of the cylinder (not to the circular ends).

Assume that the two layers of steel and the layer of glass are so thin that they present negligible resistance to heat loss relative to the resistance in the insulation layer. Thus, you could assume that the temperature at the inner surface of the insulation layer is approximately equal to the water temperature T_w. That is the inner boundary condition.

A realistic boundary condition at the outer surface of the insulation would be more complicated. For example, heat loss from the outer surface would go as

$$Q = hA(T_o - T_e),$$

where:

Q = total heat loss rate from the outside of the insulation [J s^{-1}]
h = convective heat loss coefficient for heat loss to the air
      plus thermal radiation to the surroundings [J cm^{-2} s^{-1} deg^{-1}]
A = Surface area available for heat loss [cm^{-2}]
T_o = temperature at the outside surface of the insulation [deg C]
T_e = environmental temperature [deg C]

In the steady state, this Q must be matched by conduction of heat through the insulation at its outer radius, R_o, which equals

$$Q_R = -kA\frac{dT}{dr}\bigg)_{R_o},$$

where k is the thermal conductivity of the insulation [J cm^{-1} s^{-1} deg^{-1}] and $(dT/dr)_{R_o}$ is the temperature gradient [deg cm^{-1}] just outside the insulation at $r = R_o$.

Your tasks with this problem are to:

a) derive the differential equation that has as its solution the temperature profile through the insulation.

b) obtain the analytic solution for $T(r)$ in the insulation. You would find from tables like Murphy's that

$$y = C_1 + C_2 \log x$$

is the general solution to

$$xy'' + y' = 0,$$

with C_1 and C_2 being constants of integration whose values can be calculated from the boundary conditions.

c) Let $D = 50$ [cm]
   $I = 9$ [cm]
   $k = 5.2 \times 10^{-4}$ [J cm^{-1} s^{-1} deg^{-1}] for the insulation
   $T_w = 100$ [deg]
   $T_e = 20$ [deg]
   $h = 2 \times 10^{-4}$ [J cm^{-2} s^{-1} deg^{-1}].

With these numerical values, calculate the outer surface temperature of the water heater (T_o) and the total heat loss rate, Q. Be sure to give units for Q. By what fraction would heat loss decrease if insulation thickness I were increased to 12 cm?

7. When a bird's nest gets wet, some microbial decay may occur, and heat will be generated in the process. (An Australian bird, the Mallee-fowl, incubates its eggs with heat obtained by composting vegetation in its pit-shaped nest.) Suppose that occurs in a hemispherical nest, a cross section of which appears in Fig. 8.9, p. 188. Specifically, suppose heat is produced in the nest material at a rate of g J cm^{-3} s^{-1}.

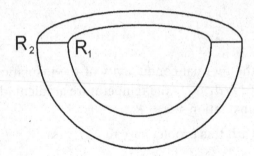

Figure 8.9: Diagram of an idealized, hemispherical Mallee-fowl nest, sectioned vertically through the middle.

Derive the differential equation whose solution is $T(r)$ for $R_1 \leq r \leq R_2$ for this situation. It may help you to know that the volume of a very thin spherical shell with inner radius r and thickness Δr is approximately $4\pi r^2 \Delta r$. Solve the equation if you can.

8. Derive and solve the differential equation describing radial diffusion of a substance in a cylinder in which the substance is partially absorbed by the medium through which it is diffusing[10]. Specifically, suppose the substance is absorbed at a constant rate of U mol m^{-3} s^{-1} in that medium. For boundary conditions, assume that at $r = 0.1$, the concentration gradient (dC/dr) is 0, and that at $r = R$, $C(R) = C_1$. (The $1/r$ factor in the solution makes it difficult to set conditions at $r = 0$.) The volume of a thin-walled cylinder of length L is approximately $2\pi r L \Delta r$. Work with concentration in mol m^{-3}.

9. Rounder Island (off the coast of Alaska) is the roundest island in the world; in fact, it is a perfect circle with a radius of $R_{outer} = 10$ km. In the center of the island, three students spend their summer internships maintaining a colony of lemmings at a constant density of $\rho_0 = 5000$ lemmings km^{-2}. This colony occupies the center circle of radius 0.1 km; that makes $\rho(0.1) = \rho_0$ one of the BCs for the problem.

The lemmings migrate outward by a diffusion-like process, with a "migrativity" of $M = 300$ km^2 da^{-1}. In particular, outward migration at any radius is given by

_____

[10]The same mathematics would apply if the substance were being lost by chemical reaction.

$$-ML\frac{d\rho}{dr},$$

where L is the perimeter of the circle at radius r, and $\rho(r)$ is the areal density of the animals at r. Assume that no births or deaths take place along the way.

Like (mythical) lemmings everywhere, whenever one of these gets to the outer rim of the island it jumps into the sea, so the density of the animals at R_{outer} is zero. Each year an equilibrium becomes established by the end of May, after which the lemming density at any given distance from the island's center ceases to change with time. (This equilibrium ends when the students go back to school in the fall.)

a) Assuming that all these conditions hold, derive the equation whose solution would tell you how the equilibrium lemming density varies with distance from the center of the island. It may be helpful to interpret the diagram in Fig. 8.5, p. 179 to represent the island.

b) Obtain the analytical (symbolic) solution, given the specified BCs.

c) Determine the numerical value of the lemming density at radius $r = 5$ km.

10. The tunnel entrance to a weasel den is shaped approximately as a circular cylinder, of radius R [cm] and length L [cm]. As the animals in the den breathe, CO_2 builds up to a stable internal concentration C_i [mmol cm^{-3}] there (mmol = millimoles). Also, the lower external CO_2 concentration is C_O [mmol cm^{-3}]. Thus, CO_2 diffuses down the tunnel from inside to outside. At the same time, respiring plant roots and microbes in the soil release S mmol of CO_2 per cm^2 of tube surface area per second into the tube. Assume that:

a) Everything reaches a steady equilibrium, so nothing changes with time;

b) L is large compared with R, so you can ignore variations of CO_2 with radius or angle at any given distance along the tunnel; and

c) The diffusivity of CO_2 in air is D [cm^2 s^{-1}].

Derive the differential equation that, if solved, would tell you how CO_2 concentration varies with distance along the tunnel. Express any quantities that depend on radius R in terms of that radius, and then simplify your equation as far as you can. Provide a unit check (and your conclusion about it) for your final equation. You need not attempt to solve the equation, but do state its boundary condition(s).

8.7 Questions and Answers

1. Is the reason for the positive sign for diffusion onto the wall of the tube just the way you defined it?

   • I'd say it's more fundamental than that. Consider both (A) the problem of stuff being carried by a bulk flow (q) of a fluid down a tube, and (B) the problem of stuff being moved by diffusion down a tube. For both problems, consider processes that remove stuff from the fluid (at rates R mg per cm of tube length per sec) and processes that add stuff to the fluid (at rates S mg per cm of tube length per sec). I'm using R for removal and S for source here.

   For problem (A), if you set up the mass balance for a short control volume of length h, then you will always get (into CV = out from CV) $qC_x + Sh \approx qC_{x+h} + Rh$.

   The two terms on the left add stuff to the box, and the two on the right subtract stuff from the box. If nothing is changing with time, then the two sides have to be equal. If you rearrange terms and take the limit, you get $dC/dx = (1/q)[S - R]$.

   For problem (B), the mass balance (based on gradients G) becomes

   $$-DAG_x + Sh \approx -DAG_{x+h} + Rh.$$

   Taking the limit here gives

   $$\frac{dG}{dx} = \frac{d^2C}{dx^2} = \frac{1}{DA}(R - S).$$

   The R and S switch signs here (compared to the bulk flow case) because of the minus sign in the "diffusion across a plane" that results from diffusion going in the direction opposite to the gradient.

2. Why was $T = a + bx$ the solution to $T'' = 0$, while $C = A \sinh ax + B \cosh ax$ was the solution to $C'' = a^2 C$?

• If I translate your question to "why are the solutions to $T'' = 0$ and $C'' = a^2 C$ so different?", I would answer that these equations have very different physical implications. The first says that T' (the temperature gradient) is the same for all x (because the *derivative* of T' is 0 everywhere). The second says that C'' is proportional to C everywhere, and thus that C' is *not* the same for all x.

To answer your original question, you can find the solution to $T'' = 0$ yourself—see Eqns. 8.13–8.14. We won't study where the sinh-cosh solution comes from (i.e., how to obtain it), but you can confirm that it is a solution (see Exercise 1, p. 183). If you're curious about how to find that solution from scratch, check in a differential equations text.

3. If you broke the animal's fur down as $\Delta r \to 0$, then couldn't you say that you were working with a sum of infinite shells, each of uniform temperature?

• Yes, that is essentially what the differential equation does. Any integration process that would be used to solve the DE then "sums" over an infinite number of infinitesimally thin shells. That takes us back to Chapter 3, where I reminded you that integration is very closely related to summation, but of continuously varying quantities rather than of discrete items.

4. If you put an impermeable cap at the end of your diffusion tube, why wouldn't that create a very large gradient across the cap?

• It's true that there might be quite a large concentration difference over some finite distance between the two sides of the cap, but what happens external to the tube is irrelevant here. The gradient that's relevant is the one-sided internal one at a point, as defined by $\lim_{h \to 0} [C(L) - C(L - h)]/h$. No locations at $x > L$ are involved here.

There would also be a zero gradient outside the tube, defined by $\lim_{h \to 0} [C(L + h) - C(L)]/h$. Those C values might be quite different from the ones inside the tube, but the gradient on the outside

also has to be zero (immediately at the surface of the cap) if there
is no diffusion there.

5. In Eqn. 8.3 on p. 167, which part represents flow from low concentration to high concentration, and which from high to low? Is C_1
the high concentration and C_2 the low concentration?

• Both versions; i.e., *both "*$+DA(C_1-C_2)/L$*" and "*$-DA(C_2-C_1)/L$*"*
represent flow from low to high concentration. Furthermore, both
versions have that property whether C_1 is higher than C_2 *or* C_2 is
higher than C_1. Try putting in some numbers (and bear in mind
that D, A, and L are all positive). If $C_1 = 100$ and $C_2 = 50$, then
both versions say that the diffusion will move mass in the positive
x direction (from C_1 to C_2). If $C_1 = 50$ and $C_2 = 100$, then both
versions tell you that diffusion will move mass in the negative x
direction (from C_2 to C_1).

6. What is W on p. 168? I don't understand deposition to the wall.
Does that mean that the molecules bond to the wall? What exactly
is adsorption?

• The process I am modelling here (probably in an oversimplified
way) is about molecules bonding to the wall, yes. That is what I
mean by adsorption. There would be a dynamic equilibrium, with
some molecules hopping on, and others hopping off. The W is
the net mass of molecules that leave the air and stick to the wall
per cm of tube length per second. A sponge absorbs water, but
molecules attaching to surfaces are said to adsorb.

7. I have a problem understanding why $W = kPC$ is positive, if that
is a way for the gas to leave the system. Since the gas is leaving,
shouldn't W be negative?

• You can define it either way. I choose to call adsorption positive (and it seems you'd like to call desorption positive). Either
would be correct, as long as you give the term the right sign in the
mass balance. In my mass balance (Eqns. 8.7 and 8.8, p. 169), I've
put "sources" to the control volume on the left side of the =, and
"losses" from the CV on the right side. If you give W the opposite
sign from my choice, you would have to put that term on the LHS.
The final result would be the same.

Chapter 9

Linear Algebra

We now leave calculus for a while, to take up methods for dealing with mathematical models that represent direct relationships among variables rather than relationships involving rates of change. First we consider *linear* models, and in Chapter 10, somewhat more complicated non-linear ones. To work with linear equations, we will use some mathematical structures—vectors and matrices—from the field of linear algebra. Near the end of the chapter, we'll apply those structures to a type of model useful in population biology, and look briefly at a variety of other applications.

9.1 Linear Algebraic Equations

The simplest linear model expresses a straight-line relationship between two variables:

$$y = b + ax.$$

Written in this form, the model suggests that for given values of x we will want to calculate corresponding values of y, and indeed the formula is often used in that way. However, for consistency with the *systems* of linear equations that are the main subject of this chapter, note that we can also write

$$ax = y - b.$$

Now it is clear that if we knew a, b, and y, we could multiply through by $1/a$, the *multiplicative inverse* of a (i.e., $1/a$), to solve for x. That is, $(1/a)ax = (1/a)(y - b)$, or $x = (y - b)/a$.

In any case, we *define* the relationship between two variables y and x to be linear if dy/dx (or dx/dy) is a constant. Put another way, this means that a change in y, which we call Δy, is proportional to a corresponding change in x (Δx).

Frequently, we are interested in systems of N linear equations in N unknowns. The general form is

$$a_{11}x_1 + a_{12}x_2 + \ldots + a_{1n}x_n = b_1$$
$$a_{21}x_1 + a_{22}x_2 + \ldots + a_{2n}x_n = b_2$$
$$\vdots \qquad \vdots \qquad \vdots \qquad \vdots \qquad \vdots$$
$$a_{n1}x_1 + a_{n2}x_2 + \ldots + a_{nn}x_n = b_n$$

while a particular example, with $N = 2$, would be Eqns. 1.1 and 1.2 from p. 2.

Here the a's and b's are usually known constants, and the x's are variables whose values are to be determined by solving the equations. If all the b's are zero, the system is termed *homogeneous*; otherwise, it is *inhomogeneous*.

9.2 How Linear Systems Arise

Linear equations arise in at least four ways:

* as natural mathematical models of real situations, often as a result of a quantity varying in proportion to one or more others

* as approximations to non-linear models, as in using the first two terms of a Taylor series

* as steps in solving other problems, including ordinary and partial differential equations, and curve fitting (regression)

* as a means of finding the equilibrium solution of a system of linear ordinary differential equations.

As an example of the first of these, suppose trout in a lake are feeding one evening on two species of insects, say of midges and moths[1]. If interested in the average caloric value supplied to the fish by each of the two insect species, we might catch two fish and examine their stomach contents. In particular, we could count head capsules to determine the numbers of each type of insect eaten by each fish, and use a bomb calorimeter to measure the total caloric content of each stomach's contents. Suppose we find:

Fish	No. of midges	No. of moths	Total Cal. in stomach
A	18	12	660
B	14	8	480

If we now let x_1 be the average caloric content of the midges eaten, and x_2 be the average for the moths eaten, we can see that:

$$18x_1 + 12x_2 = 660$$

$$14x_1 + 8x_2 = 480$$

A unit check here is pretty simple; i.e.,

$$\text{midges} \times \left(\frac{\text{cal}}{\text{midge}}\right) = \text{cal.}$$

It is easy to confirm that $x_1 = 20$ cal/midge and $x_2 = 25$ cal/moth is the solution of this pair of equations. We will study general methods for determining such solutions shortly.

Linear Equations as Approximations

That linear equations can be used to approximate more complex non-linear ones is easy to see by referring to Taylor series again. Recall from Chapter 1 that we can write any nice function as:

$$f(x) = f(a) + f'(a)(x - a) + \frac{f''(a)}{2}(x - a)^2 + \dots,$$

from which

---

[1]Modified from Chaston 1971, with permission from Elsevier. Chaston's example involves four fish and four insect species, which might give more reliable results than two and two. With more fish than insect species, one could use the statistical technique of linear regression.

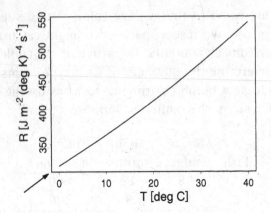

Figure 9.1: Radiative heat loss from a black body in relation to Celsius temperature. Although the energy radiated is proportional to the fourth power of the Kelvin temperature (°K=°C+273.16), the relationship is reasonably linear over the temperature range shown. This can be seen by sighting along the curve in the direction of the arrow.

$$f(x) \approx f(a) + f'(a)(x - a).$$

Truncating the series after the second term yields a linear equation that may be a tolerable approximation to $f(x)$ if we keep $|x - a|$ small enough.

As an example, remember that the Maclaurin series for e^x is

$$e^x = 1 + x + \frac{x^2}{2} + \frac{x^3}{6} + \frac{x^4}{24} + \cdots.$$

For small x, $e^x \approx 1 + x$. Thus, if $x = 0.1$, then $e^{0.1} \approx 1 + 0.1 = 1.1$. This approximation is only about 0.5% smaller than the true value, 1.105170918. For smaller x, the approximation would be even closer.

As a second example, consider the infrared radiation emitted from a black body as a function of its temperature, as shown in Fig. 9.1. This radiation is a quartic (fourth-power) function of the absolute (Kelvin or Rankine) temperature, as can be seen in the figure. However, over a temperature range of 10 degrees or so, a linear approximation is often adequate for heat-transfer calculations.

9.3 Methods for Solving Linear Systems: An Introduction to Linear Algebra

We begin this section with an introduction to the main structures used in linear algebra, namely matrices and vectors.

Matrix Notation

A *matrix* is an array of elements of the form

$$
\mathbf{A} = \begin{pmatrix}
a_{11} & a_{12} & \dots & a_{1n} \\
a_{21} & a_{22} & \dots & a_{2n} \\
\vdots & \vdots & \ddots & \vdots \\
a_{m1} & a_{m2} & \dots & a_{mn}
\end{pmatrix}
$$

The indices of the array elements come in a row-column order. That is, for elements a_{ij} with $i = 1, \dots, m$ and $j = 1, \dots, n$, the matrix has m (horizontal) rows and n (vertical) columns. To make good use of matrices, we need some further definitions and concepts:

- Two matrices, $\mathbf{A}$ and $\mathbf{B}$, are equal if and only if $a_{ij} = b_{ij}$ for all i and j.

-
$$
\mathbf{X} = \begin{pmatrix}
x_1 \\
x_2 \\
\vdots \\
x_n
\end{pmatrix} \quad \text{is a column matrix (or column vector).}
$$

- $\mathbf{T} = (t_1 \ t_2 \ \dots \ t_n)$ is a row matrix (or row vector).

- Two matrices[2] $\mathbf{A}$ (or $\mathbf{A}_{mn}$) and $\mathbf{B}$ (or $\mathbf{B}_{np}$) may be multiplied, provided that the number of columns in the first is the same as the number of rows in the second (as indicated for $\mathbf{A}$ and $\mathbf{B}$ by the common n). The product $\mathbf{C}_{mp} = A_{mn}B_{np}$ has, *by definition*, the elements

---

[2] The subscripts indicating the matrix size are often absent, but may be included when they add clarity.

$$c_{ik} \overset{\text{def}}{=} \sum_{j=1}^{n} a_{ij} b_{jk}$$

for all $i = 1, \ldots, m$ and $k = 1, \ldots, p$. For example[3], consider

$$\mathbf{A} = \begin{pmatrix} \overline{1}\ \overline{2} \\ \underline{3}\ \underline{4} \end{pmatrix} \text{ and } \mathbf{B} = \begin{pmatrix} |5\ 6| \\ |7\ 8| \end{pmatrix}.$$

The product of these matrices is

$$\mathbf{C} = \mathbf{AB} = \begin{pmatrix} (\overline{1} \cdot |5 + \overline{2} \cdot |7) & (\overline{1} \cdot 6| + \overline{2} \cdot 8|) \\ (\underline{3} \cdot |5 + \underline{4} \cdot |7) & (\underline{3} \cdot 6| + \underline{4} \cdot 8|) \end{pmatrix} = \begin{pmatrix} \overline{|19}\ \overline{22|} \\ \underline{|43}\ \underline{50|} \end{pmatrix},$$

which is of the form

$$\begin{pmatrix} a\ b \\ c\ d \end{pmatrix} \begin{pmatrix} e\ f \\ g\ h \end{pmatrix} = \begin{pmatrix} ae + bg & af + bh \\ ce + dg & cf + dh \end{pmatrix}.$$

Note that the element c_{ik} depends only on the ith row of $\mathbf{A}$ and on the kth column of $\mathbf{B}$. For example, c_{11} is formed from row 1 of $\mathbf{A}$ and column 1 of $\mathbf{B}$. As an exercise, try calculating $\mathbf{BA}$. Is it different from $\mathbf{AB}$? Why?

- Although multiplication of simple numbers is commutative (i.e., $ab = ba$), this is not generally true for matrices. For example, you should convince yourself by calculating both products for the two matrices given above that $\mathbf{BA}$ differs from $\mathbf{AB}$.

  Beyond that, it can even be true that $\mathbf{AB}$ exists while $\mathbf{BA}$ is undefined. This would occur, for example, if $\mathbf{A}$ were 3 rows × 5 columns and $\mathbf{B}$ were 5 rows × 4 columns. Then $\mathbf{AB}$ would be a 3×4 matrix, but $\mathbf{BA}$ ($[5 \times 4] \times [3 \times 5]$) would not even exist because the number of columns in $\mathbf{B}$ would differ from the number of rows in $\mathbf{A}$.

- An *identity* matrix $\mathbf{I}$, the matrix equivalent of the scalar "1", is a square matrix with ones along the "NW–SE" diagonal, and zeroes elsewhere:

$$\mathbf{I} \overset{\text{def}}{=} \begin{pmatrix} 1\ 0\ 0 \ldots 0 \\ 0\ 1\ 0 \ldots 0 \\ 0\ 0\ 1 \ldots 0 \\ \vdots\ \vdots\ \vdots\ \ddots\ \vdots \\ 0\ 0\ 0 \ldots 1 \end{pmatrix}$$

---

[3]In the example, the overbars, underlines, and vertical lines identify each number's original row or column.

This matrix gets its name because, for any matrix A, $AI=A$ and $IA=A$. You must of course use an I of the correct size to allow the multiplication—this means that I must be $n \times n$, where n is the number of columns in A for AI, or the number of rows in A for IA. I suggest you prove to yourself that this works, using a matrix like

$$A = \begin{pmatrix} 3 & 7 & 5 & 2 \\ 1 & 6 & 4 & 8 \end{pmatrix}.$$

That is, demonstrate to yourself that *pre-multiplying* A by a 2×2 identity matrix, and *post-multiplying* A by a 4×4 identity matrix both result in matrices identical to the original A.

- Multiplying a simple number a (a *scalar*) by its *multiplicative inverse* $a^{-1} = 1/a$ yields the *multiplicative identity*, 1. In a similar way, a square matrix A *may* have an inverse, denoted by A^{-1}, for which $AA^{-1} = I$. (This *defines* the inverse of a [square] matrix.) If such an inverse exists, one can show that $A^{-1}A = I$ also.

We are considering matrix notation in part because it provides a convenient shorthand for expressing and for solving systems of linear algebraic equations. Thus, if A is an $n \times n$ matrix of coefficients, X is an $n \times 1$ column vector of x variables, and B is an $n \times 1$ column vector of constants, then it is easy to confirm that

$$a_{11}x_1 + a_{12}x_1 + \ldots + a_{1n}x_n = b_1$$
$$a_{21}x_1 + a_{22}x_2 + \ldots + a_{2n}x_n = b_2$$
$$\vdots \qquad \vdots \qquad \vdots \qquad \vdots \qquad \vdots$$
$$a_{n1}x_1 + a_{n2}x_2 + \ldots + a_{nn}x_n = b_n$$

can be written as $AX=B$. For example, the trout-insect equations derived above could be written as $AX=B$ if we set

$$A = \begin{pmatrix} 18 & 12 \\ 14 & 8 \end{pmatrix}, \quad B = \begin{pmatrix} 660 \\ 480 \end{pmatrix}, \quad \text{and } X = \begin{pmatrix} x_1 \\ x_2 \end{pmatrix}.$$

As suggested above, not only does the matrix notation allow a system of equations to be written more compactly, but it can also help us to solve the system. As an illustration of the first point, note that

$$\mathbf{AX} = \begin{pmatrix} 18 & 12 \\ 14 & 8 \end{pmatrix} \begin{pmatrix} x_1 \\ x_2 \end{pmatrix} = \begin{pmatrix} 18x_1 + 12x_2 \\ 14x_1 + 8x_2 \end{pmatrix}$$

$$= \begin{pmatrix} 660 \\ 480 \end{pmatrix} = \mathbf{B}.$$

We take up the second point, using matrix algebra to solve the equations, in the next section.

Gaussian Elimination and Sensitivity Analysis

How do we solve systems of linear equations? I will describe one straightforward, relatively efficient method known as *Gaussian elimination* that could be used for hand (or spreadsheet) calculations with systems of two or three equations. (Often, other methods to be described briefly below would be applied using software like MATLAB.)

Before taking up the details of Gaussian elimination, we will consider a form of *sensitivity analysis* that is often useful with linear equations. Recall that in our fish-insect example, the right-hand sides of the equations we are solving are measurements of the caloric content of the material in the stomachs of the two fish we caught. Such measurements are never exact, of course, and it is interesting to estimate how sensitive our final answers may be to potential errors in these caloric measurements. Gaussian elimination provides an easy way to look into that issue, so we will study the solution method and sensitivity analysis simultaneously.

Suppose we believe that the calorie measurements could be in error by up to about ±10% of the correct values. Then we could solve the equations once with the real measured values, and additional times with realistic positive or negative errors added to those values. We will work with four arbitrary combinations of $594 = 0.9 \times 660$, $726 = 1.1 \times 660$, $432 = 0.9 \times 480$, and $528 = 1.1 \times 480$. Thus, we will solve the equations for five different sets of right-hand sides. This may seem like a lot of effort, but Gaussian elimination allows us to do it all at once, and as you will see, the results will be very informative.

To use Gaussian elimination for two equations with multiple RHSs, we make use of so-called *elementary row operations*, in which we may multiply any row by a constant, or may add or subtract any two rows and then eliminate one of those source rows, without changing the solution represented by the system of rows. For the present system, we first write out two rows:

		Orig.	Modified	Check
	A B	B	B's	sum
(1)	18 12	660	726 594 726 594	3330
(2)	14 8	480	432 432 528 528	2422

(9.1)

From left to right, these columns are one column of row numbers, two of coefficients from the LHS of the equations, one of the original RHSs of the equations, four of the artificially added RHSs, and a final one (the check sum) containing for each row the sum of all the other elements in that row.

Begin by driving a_{11} to 1 (Divide row 1 by a_{11}; i.e., by 18):

(1a) $1\ 0.\overline{6}^a\ 36.\overline{6}\ 40.\overline{3}\ 33\ 40.\overline{3}\ 33\ 185^b$.

Next, check to be sure that the first seven elements sum to the eighth one, which these do. Then multiply row (1a) through by 14, and subtract the resulting elements from the same-column elements in row (2):

(2a) $0\ -1.\overline{3}^c\ -33.\overline{3}\ -132.\overline{6}\ -30\ -36.\overline{6}\ 66\ -168^d$.

Divide row (2a) through by $-1.\overline{3}$:

(2b) $0\ 1\ 25\ 99.5\ 22.5\ 27.5\ -49.5\ 126$.

This row contains a great deal of information, as it really represents five separate equations, one for each RHS:

$0x_1 + 1x_2 = 25$

$0x_1 + 1x_2 = 99.5$

$$\vdots$$

$0x_1 + 1x_2 = -49.5$.

---

[a]$12/18$
[b]$3330/18$
[c]$8 - 0.\overline{6} \times 14$
[d]$2422 - 185 \times 14$

Thus, we now have, in cal per moth for the 5 RHSs,

$$x_2 \approx 25, \ 99.5, \ 22.5, \ 27.5, \ \text{and} \ -49.5.$$

These particular solutions happen to be exact, but I use the approximation sign to indicate that there will often be some round-off error.

Next, we go back to row (1a) to solve for the 5 x_1 values. E.g., for the third set of b values, we substitute the third x_2 value and solve for x_1. That is, we solve the equation

$$1x_1 + 0.\overline{6}x_2 = 33 \text{ to yield } x_1 = 33 - 0.\overline{6}(22.5) = 18.$$

We end up with 5 solution vectors, **X**, corresponding to the 5 **B** vectors, :

$$\mathbf{X} = \begin{pmatrix} 20 & -26 & 18 & 22 & 66 \\ 25 & 99.5 & 22.5 & 27.5 & -49.5 \end{pmatrix} = \begin{pmatrix} \text{cal midge}^{-1} \\ \text{cal moth}^{-1} \end{pmatrix}. \tag{9.2}$$

The fact that the solution varies so widely with relatively small ($\pm 10\%$) changes in the right-hand sides indicates that this is an *ill-determined* system of equations. We will have more to say about this later. The bottom line for a biologist should be that this is not a good way to determine the food value of moths and midges to trout!

This is a good time to note an important practical point. As far as possible, one should arrange the order of the rows (equations) to place large elements on the diagonal. This helps to ensure that no zeros occur as a_{ii} values, since each of the diagonal elements becomes a divisor at some point in the process, and more generally, to reduce round-off error. Then we apply Gaussian elimination to the rearranged system.

As noted above, software programs like MATLAB®usually use more complex methods for solving systems of linear equations, such as
L-U decomposition (Press et al. 1992). These methods can be a bit more efficient than Gaussian elimination, and can produce more accurate solutions by reducing round-off error.

There are several ways to solve linear systems with MATLAB; but each works with a matrix of coefficients **A** and a matrix (or vector) of RHSs, **B**. Here is an example for a 3×3 system, with two RHSs, that has already been arranged to put large elements on the diagonal. (MATLAB would rearrange them to optimize the row order, if we did not do so.) Define:

```
A=[7   1   3
   4  11   4
   5   2   9];
B=[4  12
  15   8
  12  16];
```

Then the command[4] X=A\B, or its equivalent X=mldivide(A,B), asks MATLAB to perform Gaussian elimination, with the result

```
X =
  -0.06041666666667   1.25833333333333
   0.96666666666667  -0.13333333333333
   1.15208333333333   1.10833333333333
```

which is of course the two solutions corresponding to the two RHSs.

The command X=B/A or *its* equivalent X=mrdivide(B,A) has the same effect for the present **A** and **B** matrices, but can produce different results from those produced by the previous commands for other matrices of different sizes. Yet another command, linsolve(A,B) produces a solution obtained by the L-U decomposition mentioned above. See help mldivide, for example, for further details of each command.

Gauss-Jordan Matrix Inversion

A second method for solving systems of linear equations involves inverting the coefficient matrix **A**, and multiplying the resulting inverse times the RHS vector **B**. This works as follows: given the equations **AX** = **B**, premultiply both sides by A^{-1} to obtain $A^{-1}AX = A^{-1}B$. (This assumes that A^{-1} exists.) Now, $A^{-1}A = I$ by definition, and because **IX** = **X**, then **X** = $A^{-1}B$. Thus, solving linear equations is formally very easy using matrix notation—the process is equivalent to solving the scalar equation $ax = b$ by multiplying both sides by a^{-1}.

In reality, however, we have to *calculate* A^{-1} to solve actual equations. There are several ways to do this, but one straightforward method is to carry Gaussian elimination a little further, with a process called Gauss-Jordan matrix inversion. Suppose we want to solve

---

[4]The backslash is deliberate, and correct.

the fish-insect equations, for example with two different right hand sides. We juxtapose the following matrices, side by side:

$A|I|B_1|B_2|\Sigma.$

Note that we add an identity matrix, of the same size as **AX**, that was not needed in Gaussian elimination.

Then we carry out a series of steps similar to those for Gaussian elimination. Here I will present all the rows of numbers first, then add the explanations below them. We have:

```
(1)  18   12  |  1     0    |  660  | 693 |  1384
(2)  14    8  |  0     1    |  480  | 528 |  1031


(1a)  1   0.6̄ᵃ |  0.05̄   0   |  36.6̄  | 38.5|  76.8̄
(2a)  0  −1.3̄ᵇ | −0.7̄    1   | −33.3̄ | −11 |  −45.4̄


(1b)  1    0ᶜ  | −0.3̄  0.45̄  |  20ᵈ   |  33  |  54.16̄
(2b)  0    1   | 0.583̄ −0.75 |  25   | 8.25 | 34.083̄.
```

The process followed to produce these rows was to:

- Write down rows (1) and (2) in the form stated above.

- Form row (1a) by dividing row (1) by its a_{11} (which was 18).

- Form row (2a) by subtracting a_{21} (i.e., 14) times the elements of row (1a), from the elements of row (2).

- Form row (2b) by dividing row (2a) by its a_{22} ($-1.\overline{3}$).

- Finally, row (1b) results from subtracting a_{1a2} (or $0.\overline{6}$) times the elements of row (2b), from the elements of row (1a).

- Be sure to check the two totals against the checksum after each new row has been calculated.

The result; i.e., rows (1b) and (2b), contains the following:

- The first two columns, which originally contained the **A** matrix, now contain the identity.

---

[a] 12/18
[b] $8 - (2/3) \times 14$
[c] $0.\overline{6} - 0.\overline{6} \times 1$
[d] $36.\overline{6} - 0.\overline{6} \times 25$

- Columns 3 and 4, which started out as **I**, now contain $\mathbf{A}^{-1}$.

- The next two columns, which originally contained the two **B** vectors, now contain the corresponding **X** vectors (solutions).

- The final column is of course the checksums, which can now be forgotten.

Compared with Gaussian elimination, we have done a little more work here to solve the two equations (with two sets of RHSs). In the process, though, we have gained the inverse of the original coefficient matrix. The advantage of this is that, if we *now* decide to solve the equations with additional **B** vectors, we can do so with relatively little work. For example, if we now have a **B** = (627, 504)′, then

$$\mathbf{X} = \mathbf{A}^{-1}\mathbf{B} = \begin{pmatrix} -0.\overline{3} & 0.4\overline{5} \\ 0.58\overline{3} & -0.75 \end{pmatrix} \begin{pmatrix} 627 \\ 504 \end{pmatrix} = \begin{pmatrix} 43 \\ -12.25 \end{pmatrix}.$$

This matrix multiplication is much easier than solving the whole system again from scratch. For each additional RHS we might want to add, we need do only a simple matrix multiplication.

To perform similar operations in MATLAB, we could enter

```
A=[18 12; 14 8]; % The first semicolon separates rows
B=[660 693; 480 528];
Ainv=inv(A)
X=Ainv*B
```

resulting in

```
Ainv =
  -0.33333333333333    0.50000000000000
   0.58333333333333   -0.75000000000000
X =
  20.00000000000006   33.00000000000006
  25.00000000000000    8.25000000000000
```

MATLAB's help page for the inv function recommends against solving linear equations in this way, however; it suggests that using X = A\B is both faster and more accurate. Either method will provide a warning for systems that are poorly conditioned (next section). However, inverse matrices can be useful for other purposes.

Solving Linear Equations Iteratively

So far we have looked at two methods for solving linear equations, methods that involve completely predictable steps for a system of any given size. There is another class of methods, called *iterative* methods, that work by repeating a series of successive approximations, until sufficient precision in the solution has been attained. We will consider one such method, the Gauss-Seidel scheme, that is advantageous in certain situations:

- For hand calculations, it may provide a solution to a system of equations more simply than other methods.

- In many applications of linear equations, the coefficient matrix may be *banded*, with non-zero elements along certain narrow diagonal bands, and zeros everywhere else. Iterative methods are often particularly efficient for systems of this type.

- Iterative methods are sometimes used to "tune up" a solution obtained by Gaussian elimination, where round-off may introduce substantial errors into the calculated solution. That Gaussian solution, however, can be used as a starting guess, and iterative methods then applied to refine its accuracy.

For two equations in two unknowns, the Gauss-Seidel method works like this. Write the two equations,

$$a_{11}x_1 + a_{12}x_2 = b_1$$

$$a_{21}x_1 + a_{22}x_2 = b_2,$$

in an order that places large elements on the diagonal as far as possible. (This is very important!) Then solve the first for x_1 and the second for x_2 (note the circularity in the two equations):

$$x_1 = \frac{b_1 - a_{12}x_2}{a_{11}} \qquad (9.3)$$

$$x_2 = \frac{b_2 - a_{21}x_1}{a_{22}}. \qquad (9.4)$$

Now

1. Guess a value for x_2 and solve Eqn. 9.3 for x_1.

2. Plug the new value of x_1 into Eqn. 9.4 and solve for x_2.

3. Check whether |New x_1 − Old x_1| < δ and |New x_2 − Old x_2| < δ, where δ is the precision required in the solutions.

4. If both conditions are met, stop; otherwise, go back to step 1 using the latest value of x_2. Continue with steps 1 to 4 until both conditions *are* met, or until the values are obviously diverging (which may occur if the coefficient matrix is not diagonally dominant).

As an example, if

$$3x_1 + 2x_2 = 10 \tag{9.5}$$

$$2x_1 + 5x_2 = 14, \tag{9.6}$$

then

$$x_1 = \frac{10 - 2x_2}{3} = \frac{2}{3}(5 - x_2)$$

$$x_2 = \frac{14 - 2x_1}{5} = \frac{2}{5}(7 - x_1).$$

Then, starting with a guess of $x_2 = 1$, the iterations yield

x_1	x_2
−	1.000
2.666	1.730
2.170	1.928
2.047	1.981
2.013	1.995
2.003	1.999
2.001	2.000
2.000	2.000
2.000	2.000.

We have successfully obtained the solution, $x_1 = 2$, $x_2 = 2$. However, note that if we had solved for x_2 from the first equation and for x_1 from the second, we would have obtained

$$x_1 = \frac{14 - 5x_2}{2} \tag{9.7}$$

$$x_2 = \frac{10 - 3x_1}{2}. \tag{9.8}$$

Here the sequence of iterations would yield

x_1	x_2
–	1.000
4.500	–1.750
11.375	–12.062
37.156	–50.734.

This sequence is diverging to $\pm\infty$, and will never converge, even though $x_1 = 2$ and $x_2 = 2$ is still the correct solution of the equations. Thus, one disadvantage of the Gauss-Seidel scheme is that it may not converge; the problem in this specific case, though, is that we have not put the large elements on the diagonal.

Finally, note that this method for solving a linear system is not limited to 2×2 systems. For example, with a 3×3 system, you solve each equation for a different one of the three x's.

9.4 Well and Ill Conditioned Linear Systems

Earlier (on p. 202) we found that with our fish-insect equations, relatively small errors of $\pm 10\%$ in the estimates of total stomach caloric content could result in very large changes in the calculated values of x_1 and x_2 (the calories/moth and calories/midge values). The problem here, in mathematical terms, is that the system of equations:

$$18x_1 + 12x_2 = 660$$

$$14x_1 + 8x_2 = 480$$

is an *ill-conditioned* (or poorly determined) system.

This ill conditioning can be seen graphically (Fig. 9.2), p. 209, if we plot x_2 against x_1 for each of the two equations. In an ill-conditioned system, the two lines representing the two equations are nearly parallel. When the RHS of one equation changes (e.g., from 660 to 693), its intercept increases and the one equation shifts upward a little. However, because the two equations are so close to parallel, their point of intersection (the solution point) shifts substantially in response.

In contrast, consider two other equations that would represent one fish with a strong preference for moths and a second fish with a strong preference for midges:

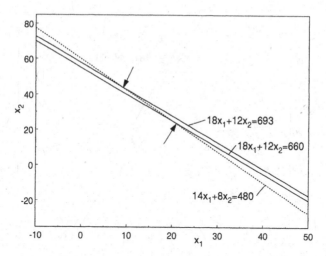

Figure 9.2: Two linear equations in a poorly determined system. A slight change in the intercept of one equation changes the solution of the system substantially. The point where the lines representing the two equations cross is the solution. Compare with Fig. 9.3, p. 210.

$$18x_1 + 4x_2 = 510 \tag{9.9}$$

$$5x_1 + 20x_2 = 425.$$

Here (Fig. 9.3, p. 210) the lines representing the two equations are far from parallel, and a small shift in one line (e.g., as 510 goes to 535) results in a much smaller shift in the solution point than before. This latter system is better conditioned than the first.

In systems of just two equations in two unknowns, graphical analyses like this are easy to do, and are quite informative. With $N > 2$ equations, graphical analysis becomes difficult because the plots must be done in N dimensions, or else all combinations of x_i-x_j plots must be produced. In such cases, a sensitivity analysis using multiple RHSs is very useful. However, additional useful information can be gleaned from the *determinant* of a system.

Determinants

The determinant of a matrix is denoted by vertical bars around the elements of the matrix and is, for a 2×2 matrix, defined as

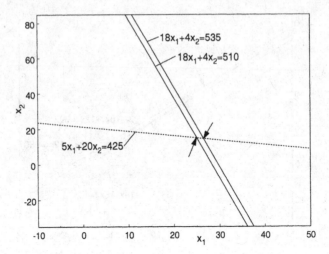

Figure 9.3: Two linear equations in a well determined system. A slight change in the intercept of one equation causes little change in the solution of the system. Compare with Fig. 9.2.

$$\text{Det}(\mathbf{A}) \overset{\text{def}}{=} \begin{vmatrix} a_{11} & a_{12} \\ a_{21} & a_{22} \end{vmatrix} \overset{\text{def}}{=} a_{11}a_{22} - a_{12}a_{21}.$$

For example, for the original ill-determined system,

$$\begin{vmatrix} 18 & 12 \\ 14 & 8 \end{vmatrix} = 18(8) - 12(14) = -24.$$

(The sign is not important for our purposes.) The value 24 is of the same order of magnitude as the larger elements in the **A** matrix, indicating that that system is poorly determined. On the other hand, for the well-determined system of Eqn. 9.9, the determinant is

$$\begin{vmatrix} 18 & 4 \\ 5 & 20 \end{vmatrix} = 18(20) - 4(5) = 340 \approx 20^2,$$

a much larger value that is similar in order of magnitude to a typical element squared.

In general, it is difficult to say exactly how large a determinant should be for a system to be well determined. However, determinants of the same order of magnitude as the absolute value of the typical elements in **A** (or smaller) indicate trouble, while for an $N \times N$ system, larger determinants on the order of the Nth power of a typical element are desirable.

The general definition of determinants for $N \times N$ matrices can be found in many books on algebra or numerical analysis. Here I will present one way to evaluate the determinant for a 3×3 matrix, and then perform the calculation for an example. The calculations shown result from expanding the determinant around the first row; more generally, any row or column can be used as a basis for expansion.

$$\begin{vmatrix} a_{11} & a_{12} & a_{13} \\ a_{21} & a_{22} & a_{23} \\ a_{31} & a_{32} & a_{33} \end{vmatrix} = (-1)^{1+1} a_{11} \begin{vmatrix} a_{22} & a_{23} \\ a_{32} & a_{33} \end{vmatrix} +$$

$$(-1)^{1+2} a_{12} \begin{vmatrix} a_{21} & a_{23} \\ a_{31} & a_{33} \end{vmatrix} + (-1)^{1+3} a_{13} \begin{vmatrix} a_{21} & a_{22} \\ a_{31} & a_{32} \end{vmatrix}.$$

For example,

$$\begin{vmatrix} -3 & 9 & 7 \\ 6 & 4 & -2 \\ 8 & -5 & 1 \end{vmatrix} =$$

$$(-1)^2(-3) \begin{vmatrix} 4 & -2 \\ -5 & 1 \end{vmatrix} + (-1)^3(9) \begin{vmatrix} 6 & -2 \\ 8 & 1 \end{vmatrix} + (-1)^4(7) \begin{vmatrix} 6 & 4 \\ 8 & -5 \end{vmatrix}$$

$$= -3[(4)(1) - (-2)(-5)] - 9[(6)(1) - (-2)(8)]$$

$$+ 7[(6)(-5) - (4)(8)] = -614.$$

Because 614 is not too different from the largest element taken to the Nth power ($9^3 = 729$), a system of equations with these coefficients would probably be reasonably well conditioned. A sensitivity analysis would be useful here.

As a further note, recall (perhaps from your distant memory) that if one equation in a system is a constant multiple of the other, then the system cannot be solved. For example, if all one has is the two equations:

$$x + 2y = 5$$

$$3x + 6y = 15,$$

then x and y cannot be determined at all. For this system, the determinant is $(1)(6) - (2)(3) = 0$. Graphically, these "two equations" determine the same line in the x-y plane. Thus, they do not intersect at a unique point, and this is consistent with the indeterminacy of x and y. In the same vein, it is an interesting exercise to attempt to invert

$$A = \begin{pmatrix} 1 & 2 \\ 3 & 6 \end{pmatrix}$$

using the Gauss-Jordan procedure.

Preparation for the Next Chapter

Before turning to the next chapter, try to figure out some practical way to estimate a numerical value of x for which $e^{-x} = \sin x - \log x$, where x is in radians (not degrees) in the sine function. Your method should be able to produce a solution good to three significant digits.

9.5   Population modelling with Leslie matrices

As noted in the introduction to this chapter, matrices and their vector components have other applications beyond providing solutions to systems of linear equations. Models based on matrices are often used to model population growth, taking account of the age or life-stage structure of a population. The notation here is similar to that in Chapter 2 of Caswell (2001), where you will find much more detail. Beissinger & Westphal (1998) also discuss these models, but provide warnings about using them for predicting viability of rare and endangered populations.

The basic idea is captured by the *Leslie matrix*, which we consider for an example population with three age classes:

1. Suppose the age classes are one year "wide," i.e., the population contains individuals only in age classes 0–1 yr old, 1–2 yr old, and 2–3 yr old. (Time units of months, decades, etc. could also be used. Five-year classes are sometimes used in human demography, for example.) We will use the model to project from time t to $t + 1$, $t + 1$ to $t + 2$, etc. In some applications, the categories can be life stages (such as larvae, pupae, and adults in an insect population) rather than age classes.

2. The state of the population at any time t is represented by a vector $\mathbf{n}(t)$ that has three elements, $n_1(t)$, $n_2(t)$, and $n_3(t)$ that give the number of individuals in each category at that time.

3. Some fraction of the critters in Category 1 mature and move to Category 2 in a year, and a (probably different) fraction move from 2 to 3. In this example population (which might represent an insect species), those in the oldest age class all die, so none move into a fourth or higher age class. This process is represented by $n_2(t+1) = P_1 \times n_1(t)$ and $n_3(t+1) = P_2 \times n_2(t)$, where P_i is the probability of an individual in the ith age class surviving for a year and moving into the next higher class[5].

4. $n_1(t+1)$ can't be determined using that same approach, because the new members of the first class come from births, not from surviving from an earlier class. Rather, this value is modelled as $n_1(t+1) = F_1 \times n_1(t) + F_2 \times n_2(t) + F_3 \times n_3(t)$, where the F values are the per-capita fertilities of each age class (or other category).

All those relationships can be summarized with matrices, *viz.*,

$$\begin{pmatrix} n_1 \\ n_2 \\ n_3 \end{pmatrix}(t+1) = \begin{pmatrix} F_1 & F_2 & F_3 \\ P_1 & 0 & 0 \\ 0 & P2 & 0 \end{pmatrix} \begin{pmatrix} n_1 \\ n_2 \\ n_3 \end{pmatrix}(t),$$

which using matrix notation becomes

$$\mathbf{n}(t+1) = \mathbf{A}\mathbf{n}(t).$$

Caswell calls $\mathbf{A}$ a *population projection matrix*[6], and points out that its elements can be constants (perhaps unrealistically), functions of environmental changes and hence of time, functions of population size (indicating density dependence), or combinations of the latter two.

---

[5]These are examples of *finite difference* equations; note their similarity to the first stage of deriving mass balance differential equations by the $m(t+h) \approx \dots$ approach we have studied. For these, however, we keep a finite time step, and do not take a limit.

[6]This form is often termed a Leslie matrix in other writings.

Example: Suppose we are modelling a population for which

$$A = \begin{pmatrix} 0 & 2 & 7 \\ 0.2 & 0 & 0 \\ 0 & 0.45 & 0 \end{pmatrix} \text{ and } n(0) = \begin{pmatrix} 100 \\ 20 \\ 5 \end{pmatrix}. \tag{9.10}$$

Then

$$n(1) = \begin{pmatrix} 0 \cdot 100 + 2 \cdot 20 + 7 \cdot 5 \\ 0.2 \cdot 100 + 0 \cdot 20 + 0 \cdot 5 \\ 0 \cdot 100 + 0.45 \cdot 20 + 0 \cdot 5 \end{pmatrix} = \begin{pmatrix} 75 \\ 20 \\ 9 \end{pmatrix}.$$

The reader is left to show that $n(2) = [103\ 15\ 9]'$, and $n(3) = [93\ 20.6\ 6.75]'$; these vectors represent the evolving population structure in years 2, and 3. As you work, think about the meaning of each individual element of the A matrix. For example, what is the meaning of the 2 in the first row, and of the 0.45 in the third row?

Note that this process is made easier by software like MATLAB, which gets its name from its original purpose of working with matrices. Even spreadsheets have functions allowing for matrix multiplication, although these functions are often not very "transparent."

Eigenvalues

Here we look briefly at the concept of *eigenvalues*[7] of matrices, which arise in many areas of science and math in addition to population modelling. By definition (Derrick & Grossman 1981), an $n \times n$ matrix A has an eigenvalue λ if there is a non-zero vector[8] v such that

$$Av = \lambda v.$$

(It may have up to n of them.)

One can find eigenvalues from the *characteristic equation* of a matrix, which results from setting the determinant of $A - \lambda I$ equal to zero. For example, for

$$A = \begin{pmatrix} 3 & -1 \\ 2 & -5 \end{pmatrix}, \ A - \lambda I = \begin{pmatrix} 3 - \lambda & -1 \\ 2 & -5 - \lambda \end{pmatrix},$$

---

[7]"Eigen" is German for "own."

[8]Both the eigenvalue and the components of the eigenvector can be real or complex numbers.

so the characteristic equation is

$$(3 - \lambda)(-5 - \lambda) - (-1)(2) = 0.$$

For a 2×2 matrix, that will be a quadratic, which here has a solution of $-1 \pm \sqrt{14}$. Hence, those are the two eigenvalues for that matrix. (These days, software like MATLAB is generally used to find eigenvalues; e.g., with the function `eig(A)`.)

Two examples of how these are used are

1. With Leslie matrices, the dominant (largest) eigenvalue gives the geometric rate of population change. For example, for the **A** of Eqn. 9.10, the dominant eigenvalue is 1.0114. This indicates that the modelled population will grow with time according to $N = N_0 \exp(1.0114t)$, where N is the total population size.

2. With systems of linear ordinary differential equations like Eqns. 6.1 on p. 137, the eigenvalues of the matrix of coefficients provide the rate constants in the exponential functions that appear in the solution of the system.

Eigenvalues also have uses in multivariate statistics, for example in principle component analysis.

9.6 Other Applications for Matrices

Matrices have many other applications beyond helping to solve systems of linear equations and modelling population growth. They are often used in some areas of statistics, for example. In multiple regression, the least-squares estimates **b** for the vector of regression coefficients can be represented and calculated as $\mathbf{b} = (\mathbf{X'X})^{-1}\mathbf{X'y}$, where **X** is the matrix of independent (causative) variables, and **y** is the vector of responses (Greene 1993).

Another type of matrix application is the "Markov-chain" model, in which entities existing in a set of conditions move (or not) to other conditions, with transition probabilities given by a matrix similar to a Leslie matrix. Example applications include modelling:

- forest succession, with transitions of trees in a forest from species to species over time (Horn 1976);

- dispersal of organisms among patches (Caswell 2001);

- growth of pine trees from smaller girth classes to larger ones (Mesterton-Gibbons 1995);

- sequences of weather types in a city (Grossman and Turner 1974); and

- passage of genetic traits from generation to generation (Maki and Thompson 1973). These authors also describe the theory of Markov chains in some detail.

9.7 Exercises

1. Solve the following set of linear algebraic equations using Gaussian elimination.

$$22.7x - 11.4y = 13.76$$

$$10.3x + 20.2y = 6.41$$

Carry at least 7 figures in your calculations. State whether or not you think this system is highly sensitive to small changes in the right-hand sides. On what did you base your opinion?

2. Consider the set of linear equations given by $AX = B$, where

$$A = \begin{bmatrix} 1.7 & 2.3 & 4.7 \\ 3.9 & 1.2 & -2.1 \\ -3.0 & 5.1 & 3.0 \end{bmatrix}.$$

There are three different B vectors; i.e.,

$$B_1 = \begin{bmatrix} 10.5 \\ 3.0 \\ 4.9 \end{bmatrix}, \quad B_2 = \begin{bmatrix} 11.03 \\ 2.85 \\ 4.90 \end{bmatrix}, \text{ and } \quad B_3 = \begin{bmatrix} 10.50 \\ 3.00 \\ 5.15 \end{bmatrix}.$$

Note that the second and third Bs include small perturbations of the elements in the first one.

A. Solve these equations using Gaussian elimination, including all three B vectors in your solution. To do this, write the coefficient

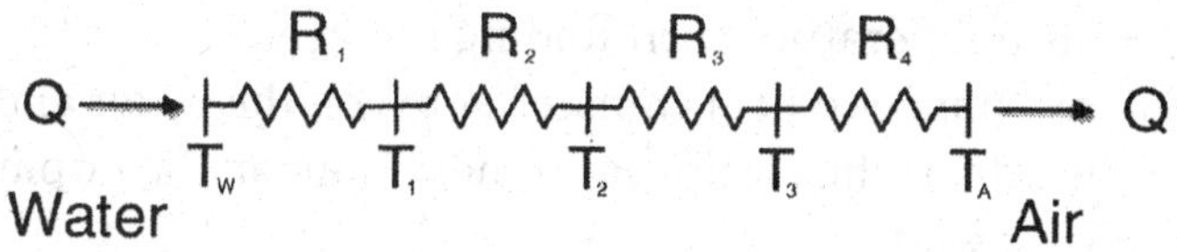

Figure 9.4: Thermal resistor model for loss of heat through the walls and insulation of a water heater

matrix together with the three **B** vectors and with a column containing the sum of all numbers in each row:

A matrix			three Bs			check sum
1.7	2.3	4.7	10.5	11.03	10.50	40.73
3.9	1.2	-2.1	3.0	2.85	3.00	11.85
-3.0	5.1	3.0	4.9	4.90	5.15	20.05

The check-sum column allows you to check your calculations at each stage. At any step in the Gaussian elimination process, the last element of each row should equal the sum of all other elements. Be sure to rearrange the equations to put the largest elements on the diagonal, as far as possible. Comment on the sensitivity of the solution to the given 5% changes in the **B** vector.

B. Check your results by substituting back into the original equations.

3. A company manufacturing hot water heaters wants to add an extra layer of insulation to their units, in response to consumer demands as gas prices rise. The present line of heaters uses 5 cm of Fluff brand insulation, which is pretty expensive but can stand temperatures well above $100^{0}C$. Another brand (Airy insulation) is available at less than half the cost, but it deteriorates abruptly if it gets hotter than $78^{0}C$. The question is, can a 5 cm layer of Airy be added outside the 5 cm of Fluff, without the Airy's getting too hot? (Airy has the same heat transfer properties as Fluff has.)

The following theory applies; see Fig. 9.4.

Let T_w = water temperature (all temperatures in degrees Celsius)
  T_1 = temperature at the inside of the Fluff
  T_2 = temperature at the Fluff-Airy interface
  T_3 = temperature at the outside of the Airy

T_A = air temperature surrounding the heater

R_1 = resistance to heat transfer between the water and the inside of the Fluff layer (inner boundary layer plus tank wall)

R_2 = resistance to heat transfer through the layer of Fluff

R_3 = resistance to heat transfer through the layer of Airy

R_4 = resistance to heat transfer between the outside of the Airy and the air (heater wall plus outer boundary layer)

By the law of conservation of energy, the total heat loss rate Q (per unit surface area) must be the same through each layer. Also, theory tells us that

$$Q = \frac{T_W - T_1}{R_1} = \frac{T_1 - T_2}{R_2} = \frac{T_2 - T_3}{R_3} = \frac{T_3 - T_A}{R_4}.$$

because of the way the resistances are defined.

Since all the fractions above are equal to the same heat loss rate (Q), you can form three equations in the three unknowns (T_1, T_2, T_3), and Q need not be known to solve for the temperatures. In forming your equations to eliminate Q, be sure to use each of the four fractions above at least once.

Suppose the thermal properties of the two kinds of insulation lead to

$$R_1 = \frac{18°C}{W}, \ R_2 = R_3 = \frac{12,000°C}{W}, \ \text{and } R_4 = \frac{880°C}{W}.$$

Finally, then, choose a worst case and calculate the resulting temperature on the inside of the Airy insulation. In particular, the water may get as hot as $T_w = 100°C$ maximum. On a hot summer's day, the air temperature surrounding the heater might go to $T_A = 45°C$. ("W" is the standard abbreviation for "Watt.")

Once you've derived the necessary equations, solve them either by Gaussian elimination (e.g., in a spreadsheet), or using MATLAB.

A. For these values, what is T_2? Can we use the Airy, if its properties are as stated?

B. How sensitive is the calculated value of T_2 to the estimates of T_W and T_A? In particular, if $T_W = 102.5°C$, what would T_2 be? Could we use the Airy in this case?

Two points of interest:

i. The general principles used above apply to a wide range of problems involving either heat transfer or mass diffusion through a series of layers.

ii. Actually, I have approximated the problem to allow for an exact answer. More correctly, one would have to adjust the resistances to allow for the curvature of the layers. I expect the error is not large, however.

4. For a study of temperature in birds' nests, you borrow an old instrument that uses thermistors to sense temperature. The manual for the instrument tells you that the resistance rises nearly linearly with the temperature, over a temperature range between 0–100° C, and that you should calibrate the system every six months or so to account for any aging of components. So, you obtain output readings of r_1 kO (kilohms) at T_1 °C and of r_2 kO at T_2 °C.

A. Find the equation (in terms of the symbols r_1, r_2, T_1, and T_2) that would give you temperature as a function of resistance in future measurements.

B. If r_1 = 8.7 kO and r_2 = 12.5 kO when T_1 = 0° C and T_2 = 40° C respectively, what would the sensor temperature be when the reading is 9.2 kO?

Hint: Of the various formulas for working with straight lines, the "two-point form" is often the most efficient. If you know two specific points (x_1, y_1) and (x_2, y_2) on a straight line, and if (x, y) is any other point on the line, then the line can be defined by:

$$\frac{y - y_1}{x - x_1} = \frac{y_2 - y_1}{x_2 - x_1}.$$

This works because both sides of the equality are expressions for the slope of the line. You may use this if you like, but are not required to do so.

5. The "equilibrium solution" to a system of differential equations is the set of values from which there is no further change. At that point, because none of the dependent variables is changing, all the derivatives are equal to zero. Thus,

$$\frac{dx}{dt} = f_1(x, y, z, t) = 0$$

$$\frac{dy}{dt} = f_2(x, y, z, t) = 0$$

$$\frac{dz}{dt} = f_3(x, y, z, t) = 0$$

becomes a set of regular (not differential) equations of the form $f_1 = 0$, $f_2 = 0$, $f_3 = 0$ at the equilibrium point x_e, y_e, z_e.

From about 1965 to 1975, ecologists were exploring models that represented ecological communities as systems of interacting populations that could be described by systems of linear differential equations. A simple example of such a model would be

$$\frac{dR}{dt} = 0.100\,R - 0.900\,F - 30$$

$$\frac{dF}{dt} = 0.003\,R - 0.020\,F - 20$$

where R denotes the number of rabbits, and F the number of foxes, occupying some area of land.

If this were an accurate model (a very big IF), then for what number of rabbits and what number of foxes would their populations be stable (i.e. not changing with time)? Use Gaussian elimination or MATLAB to find your solution.

6. Finite elements and finite differences represent two approaches for solving partial differential equations numerically (and approximately). Both these numerical approaches tend to yield large systems of linear equations, of which a small example is provided here. Like the system here, the equations involved usually have many zero coefficients, and this can simplify their solution. Consider this example, which applies to heat transfer along an object that projects out into a moving fluid (like a tree branch or a cooling fin):

$$
\begin{array}{rrrrcr}
20\,T_1 & -17\,T_2 & +0\,T_3 & +0\,T_4 & = & 150 \\
-17\,T_1 & +40\,T_2 & -17\,T_3 & +0\,T_4 & = & 120 \\
0\,T_1 & -17\,T_2 & +40\,T_3 & -17\,T_4 & = & 120 \\
0\,T_1 & +0\,T_2 & -17\,T_3 & +21\,T_4 & = & 84.
\end{array}
$$

As you know, if you were to use Gaussian elimination to solve these equations, you would end up with an array of elements of the form just below, and as you can see, the system above already has three of the desired zeros in the lower left corner.

$$
\begin{array}{ccccc}
1 & a_{12} & a_{13} & a_{14} & b_1 \\
0 & 1 & a_{23} & a_{24} & b_2 \\
0 & 0 & 1 & a_{34} & b_3 \\
0 & 0 & 0 & 1 & b_4
\end{array}
$$

Your task here is to *begin* a Gaussian elimination, and take it through the first two major steps. That is, set up the Gaussian elimination, and carry out the appropriate computations to the point where the first two columns are of the form

$$
\begin{array}{ccc}
1 & a_{12} & \cdots \\
0 & 1 & \cdots \\
0 & 0 & \cdots \\
0 & 0 & \cdots
\end{array}
$$

For this exercise, you need not complete the computations, nor perform any back substitution (but do show all the other columns, of course). The idea is to demonstrate that you have the basic idea of Gaussian elimination. Keeping a check sum column is optional, but you omit it at your own peril.

7. You are working with an engineer in testing a three-reactor system designed to reduce the amount of oxygen-demanding wastes (ODWs) in a wastewater treatment plant. You have derived three ODEs as a model of that system, namely:

$$
\frac{dC_1}{dt} = \frac{q_i C_i - q C_1 + q_f C_3}{V_1} - f_1 C_1
$$

$$
\frac{dC_2}{dt} = \frac{q C_1 - q C_2}{V_2} - f_2 C_2
$$

$$
\frac{dC_3}{dt} = \frac{q C_2 - q C_3}{V_3} - f_3 C_3
$$

where C_1 to C_3 are the concentrations of ODWs in the three tanks [g m^{-3}], V_1 to V_3 are the tank volumes [m^3], q_i and C_i are the flow rate [m^3 hr^{-1}] and ODW concentration of the wastewater inflow to the system, q_f is the rate at which water is fed back from the third tank to the first, to "seed" the system with active bacteria that aid in ODW degradation. The sum $q = q_i + q_f$ represents the flow between tanks 1-2 and 2-3. Finally, f_1, f_2, and f_3 [hr^{-1}] are the fractions of ODW destroyed or settled out per hour in the three tanks.

Suppose that q_i and C_i both remain constant for an extended test period. Then the three tank concentrations would approach equilibrium levels that can be estimated by setting the three derivatives to zero and solving the resulting algebraic equations. Let $q_i = 150$ m^3 hr^{-1}, $q_f = 10$ m^3 hr^{-1}, $C_i = 120$ g m^{-3}, $V_1 = V_2 = V_3 = 80$ m^3, $f_1 = 0.8$, $f_2 = 0.7$, and $f_3 = 0.6$.

   a) Set up the **A** matrix and **B** vector in terms of those coefficient names.

   b) Determine the numerical values of the elements of **A** and **B**.

   c) Use MATLAB or other software to solve this system. Interpret the results, being sure to state units.

8. A large Midwestern power plant burns a mixture of high-sulfur Indiana Coal (4% sulfur by weight) and low-sulfur Wyoming coal (1.5% sulfur). The western anthracite has higher energy content ($W = 2500$ BTU/ton) than the Indiana bituminous coal ($I = 2100$ BTU/ton). (Note: these numerical values for W and I are just rough estimates, but the concept is reasonable.)

For each 100,000 BTU produced, the plant is allowed to emit no more than 1 ton of sulfur. Unfortunately, because the western coal is considerably more expensive than the Midwestern, the plant manager chooses to emit that full amount. How many tons of each type of coal should the plant burn for each 100,000 BTU of heat energy? Use Gaussian elimination or software to determine the unknown tonnages. Provide units for all quantities.

9.8 Questions and Answers

1. Is the fish-insect system so sensitive because there are only two equations? Would the system become less sensitive if there were more equations, or more unknowns, or both?

• The number of equations isn't the issue. For example, the system $1x + 0y = 5$ and $0y + 1y = 8$ would be perfectly well determined. If you changed the 5 by 10%, then x would also change by 10%. The problem with the original trout-insect equations (which we look at in more detail later in the chapter) is that they are not very different—the two fish ate similar proportions of the two insects. The system would be a lot less sensitive if Fish 1 had eaten mostly midges and Fish 2 had eaten mostly moths. Section 9.4 explains this phenomenon geometrically.

2. Please give an example of an overdetermined system.

• If we had data for a third fish, say one that had eaten 5 midges and 12 moths, but there were still only two insect types, then we would have three equations in two unknowns. That third equation would not be a linear combination of the first two, either, because of the numbers I picked. Thus, the three equations would not have a unique solution. A good solution in such a case would be to estimate the average caloric content of the two insect types by regression, a statistical procedure.

3. Please explain again why you can sometimes multiply **AB**, but not **BA**, with the same two matrices.

• The best way to understand this is probably to take an example, and try to carry out the multiplication. Then you'll see the *why* of it. For example, try to calculate $C = AB$ and $D = BA$ when

$$A = \begin{pmatrix} 1 & 2 & 3 \\ 5 & 6 & 7 \end{pmatrix} \quad \text{and} \quad B = \begin{pmatrix} 1 & 2 & 3 & 4 \\ 5 & 6 & 7 & 8 \\ 8 & 7 & 6 & 5 \end{pmatrix}.$$

C should have two rows and four columns. What do you get for **D**?

4. Where do the check sums come from? Are they given in the problem, or do you have to figure them out? If so, how?

• Take the fish-insect example (because the numbers there are simple).

$$18x_1 + 12x_2 = 660$$

$$14x_1 + 8x_2 = 480.$$

There are no check sums inherent in the problem. When you set up the matrices to solve this system by Gaussian elimination, you could write

$$18 \ 12 \ 660$$
$$14 \ \ 8 \ 480$$

and just work with those. Instead, it's helpful to add $18 + 12 + 660 = 690$ to the end of the first row, and $14 + 8 + 480 = 502$ to the end of the second row.

Now, when I divide the first row through by 18, I get 1, 0.66666, 36.66666, and 38.33333. Then I check whether $(1 + 0.66666 + 36.66666) = 38.33333$, which it does. In other words, at every step after the first writing down, there are two ways of arriving at the check sum (once from the row operations, and once from summing), and they should both give you the same result.

That's the check. If you check at every step in this way, you save yourself from propagating an error from one step to another, and having to redo the whole analysis when your results don't check at the end. (Be sure you also know how to, and *do*, check your results by substituting them back into the original equations.)

5. I'm still not grasping what the solutions to these equations mean. What are they telling us? What situations would occur where I should know to solve the equations by Gaussian elimination or by matrix inversion?

• I wonder whether this is one of those things that is so easy it seems hard? All kinds of situations lead to linear equations because some response changes in direct proportion to one or more "causes." You'll see several examples in the chapter exercises. In fact, if you turn back to § 1.1 of this book, you'll see that we started with such an example. The question "How much each of 90% alcohol and 40% alcohol must be mixed to produce 1 liter of solution that is 2/3 alcohol?" resulted in the two equations

$V_1 + V_2 = V_3$ (total liters must add up) and

$f_1 V_1 + f_2 V_2 = f_3 V_3$ (total alcohol must add up).

You should recognize these as two linear equations in two unknowns. If you have just two, it's not hard to solve them analytically (symbolically) as we did with those. However, if you had three linear equations in three unknowns, it becomes far too cumbersome to get an analytic solution, and then you would turn to one of the methods we've been studying.

The solutions we obtain are the things we want to know (the "unknowns"). In the example just given, the solution (mathematically speaking) is the amounts of the two mixtures we would need to mix to get the mixture of 2/3 alcohol that we need in the lab. In the water heater exercise, the x values are the temperatures at various locations in the water heater. In Exercise 5 of this chapter, the solution tells us the numbers of rabbits and foxes that would exist if this modelled "ecosystem" ever reached a steady, unchanging condition.

6. What situations would occur where I should know to solve the equations by Gaussian elimination or by matrix inversion?

• If you are pretty sure that you already have all the RHSs for which you will want solutions, then use Gaussian elimination because of its greater efficiency. On the other hand, if you think you may later want to solve the same equations with additional RHSs, then matrix inversion is probably the better choice.

7. What does it mean for a system of linear equations to be sensitive or not?

• If a system is ill-conditioned (sensitive), that means that you really can't trust your solution unless you know that you've measured all the constants very accurately. Small errors in either the a constants or the b constants can be "magnified" into large errors in the x results. For example, in the alcohol mixture problem, if the system were ill-conditioned, you would want to be more careful than usual to measure amounts accurately, and to be sure about the percentage alcohol contents, too. If a system is well-conditioned, then small errors in the inputs (a and b values) lead to similarly small errors in the x results.

8. I still have a hard time understanding why arranging large elements on the diagonal produces less variation in the answers (i.e., if you change them by 10%) than if you don't so arrange them.

• I don't think that's correct. Putting large elements on the diagonal is a related but different issue from the sensitivity (degree of determination) of the equations themselves. It's true that if there *are* dominant large elements that can be put on the diagonal, then the system will be well-determined. Beyond that, though, having large elements on the diagonal reduces round-off error in the calculations for Gaussian elimination or for matrix inversion. To see this, try solving

$$1000x + 1.5y = 1001$$

$$41.5x + 1000y = 1001$$

in that order, and then solve in the reversed order,

$$1.5x + 1000y = 1001$$

$$1000x + 1.5y = 1001.$$

In both cases, round off to four significant digits at every step. The correct solution is $x = y = 0.9995007489$ (if you carry lots of digits). If I round to four digits at every step with the large elements on the diagonal, I get $x = y = 0.9995$ (not bad). If I round to four digits with the equations arranged in the second way, I get $x = 1$ and $y = 0.9994$ (not as good). Note that these equations are very well determined—the determinant is very close to the square of the largest element.

With larger $N \times N$ systems, you do more calculations, so round-off errors can build up even more than in a 2×2 system. *This* is the reason for putting large elements on the diagonal.

9. In the Gauss-Seidel (iteration) method for solving systems of linear equations, how do you determine δ?

   • The precision needed in the solutions depends on your application. In the fish-insect application, we might want to know average calories per midge to the nearest 0.1 calorie. That would be δ. In the exercise on p. 219, we might want to know the numbers of rabbits and foxes to the nearest whole animal. In that case, $\delta = 1.0$. Choosing δ requires knowledge of the situation, and judgment—it is not a purely mathematical process.

10. What is the sum column in Gaussian elimination used for, other than for checking?

• That's its only purpose, but it is well worth doing. If you ever solve a 3×3 system without the checks, and then discover at the end that your 'x' values don't satisfy the original equations, you will wish you had caught your error (or errors) sooner.

11. In Eqns. 9.7–9.8, p. 207, you show how the Gauss-Seidel scheme fails when the large elements are off the diagonal. Can you please show how the equations were aligned to produce this result?

• Yes. This results from solving Eqn. 9.5 for x_2 instead of x_1, and solving 9.6 for x_1 instead of x_2.

12. In the Markov-Chain succession model, how are the probabilities calculated?

• Broadly, Horn estimated them from tree-density data in forest stands of different successional ages. If you want more detail, see Horn's paper. The reference is on p. 265.

13. Would that model yield more accurate results if you used more, shorter time steps?

• I suppose it would, but the probabilities would have to be determined for that shorter time step. The shorter the time step, the smaller each element in the P matrix would be.

14. Please explain again why the largest numbers must be on the diagonal.

• I don't know whether you are asking this in the context of Gaussian elimination, or matrix inversion, or Gauss-Seidel, but I guess the reason is generally the same for all three. In each of those processes, you end up dividing by the diagonal elements. If those divisors are big, the division tends to 'damp out' errors, but if they are small, then any errors in the system (including roundoff errors) tend to be magnified. It's clear you can't have *zeros* on the diagonal, especially, because of the division.

15. Why does each row in Eqn. 9.1 on p. 201 average to the first term in each?

• Funny you should notice that. It hadn't occurred to me, but it is explained by the B elements in Eqn. 9.2 being the measured caloric value, and plus or minus 10% around the measured value.

Thus, the B values average to the measured value. Since the x's are linear functions of the b's, the x's must average to the value for the average b. It is a general feature of linear relationships that the mean of function values is the same as the function value at the mean. That's *not* true for non-linear relationships.

16. How does Gaussian elimination relate to statistics, if at all? It seems that the example we did in class should relate to the standard deviation of cal/midge, etc.

    • There might be some sort of relationship for our *estimate* of cal/midge, since we've worked with estimated measurement error. I don't think there would be a relationship with the standard deviation of the caloric content of various individual midges, though. Finding slopes and intercepts in linear regression (and multiple linear regression) involves solving systems of linear equations, and Gaussian elimination might be used by some statistical software to do that. That's the only sort of relationship I can see, however.

17. Is Gaussian elimination the method used to solve a larger number of simultaneous equations, say for ten variables?

    • Often, yes. When I have some 800 simultaneous equations as a step in solving a partial DE for CO_2 diffusion inside a leaf, that's the method I use, because of its efficiency. Other methods are sometimes used for certain applications, however. For example, some numerical processes lead to systems that have non-zero elements on only a narrow band centered on the main diagonal. Sometimes iterative solution methods (similar to the one in Sect. 9.6) are useful there. Also, a method called L-U decomposition is popular these days—I think it is good for reducing round-off error.

18. What does the expression

$$c_{ik} = \sum_{j=1}^{N} (a_{ij} \cdot b_{jk})$$

mean? Why j=1?

    • The capital sigma means to take the sum of the expression to the right, as the index j varies over the integers from 1 to N. In our 2×2 times 2×2 matrix example, N is 2. For each combination of i=1 or 2 and k=1 or 2, j varies from 1 to 2.

Chapter 10

Non-Linear Equations

In Chapter 9, we studied linear equations as models for real phenomena and as approximations to non-linear relationships. In many cases, however, the true relationships between or among variables are too far from linear to allow useful linear approximation, so now we take up methods for dealing with non-linear relationships. A non-linear equation is one that represents a relationship whose graph is not a straight line (with two variables), a plane (with three), or a hyperplane (with four or more). Examples, from a wide range of possibilities, include:

- Polynomials like quadratics, cubics, quartics, etc.

- Exponentials (ae^{bx}, ae^{-bx}) and the related $\sinh x$ and $\cosh x$.

- Logarithmic and trigonometric functions.

- Power functions, like ax^b.

- Many others (e^{-ax^2}, etc.)

Often these functions arise as solutions of differential equations, but they can arise in other ways too. In my view, linear functions are used too often (particularly in regression studies in statistics). Linear functions are not appropriate when:

- there is obvious curvature in a relationship.

- a response of interest oscillates.

- quantities approach asymptotes.

In other words, use common sense in choosing models to represent data.

10.1 Roots of Nonlinear Equations

Working with nonlinear equations takes various forms. Most simply, one has functions like $y = ae^{-kt} + be^{-rt}$, and needs to calculate the value of y at one or more values of t when the parameters a, b, k, and r are specified. (Given "x," find "y.") The Streeter-Phelps equations, a well-known model for oxygen loss and replacement in a stream carrying oxidizable wastes (pp. 144, 134), lead to a solution of this form, for example. Given modern calculators, computations of this type are straightforward.

A more difficult but still common problem arises when one has a nonlinear function of the form $y = f(x)$ and needs to find a value for x when the value of y is specified. For example, if $y = e^x$ and we learn somehow that $y = 3$, then $\log 3 = \log e^x = x$, so $x \approx 1.0986$. Similarly, if $y = 3x^2 + 1.5x + 0.1$ and we know that $y = 5.27$, then it is easy enough to apply the quadratic formula to determine the two corresponding values of x. But many problems can't be solved so easily. For example, if $y = x^4 + 13x^3 + 7x^2 + 3x - 5$ and $y = 17$, then there is no "quartic formula" analogous to the quadratic formula to tell us analytically what x must be. For problems like this we must once again turn to numerical methods.

Problems of this type can be formalized as follows. Suppose $y = g(x)$, and we want to know the value (or values) of x that make $y = c$, where c is some specified constant. Then we can always restate the question by writing $f(x) = g(x) - c$, and asking **"What values of x cause f(x) to equal zero?"** Such values are called *roots* or *zeros* of $f(x)$—we devote the rest of this chapter to methods for finding roots of functions. Warning—the methods we study apply *only* to finding values of the independent variable (x) that drive the function value to *zero*. Thus, you must remember to convert your problem to that form before attempting to solve it!

Note that the problem posed on p. 212; i.e., to find x such that $e^{-x} = \sin x - \log x$, is another example of this type of problem. If we write $f(x) = e^{-x} - \sin x + \log x$, then the number we seek is a root of $f(x)$ (Fig. 10.1, p. 231). Were you successful in solving that?

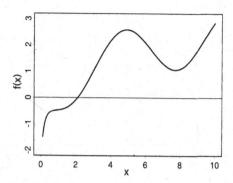

Figure 10.1: A non-linear function, $f(x) = e^{-x} - \sin x + \log x$, for which we seek a root.

We now take up a few (out of many) methods for finding roots, and apply these to an example from physiological plant ecology. The energy balance for a plant leaf whose temperature is not changing with time can be written as

$$Q_A = h_c(T - A) + \sigma \epsilon T^4 + LE,$$

where

Q_A = total solar and thermal radiation absorbed by the leaf
h_c = convective heat transfer coefficient
T, A = leaf and air Kelvin temperatures, respectively
σ = the Stephan-Boltzmann constant
ϵ = emissivity of the leaf to long-wave thermal radiation
L = latent heat of vaporization of water
E = flux density of evaporative water loss.

If the stomata (pores) of the leaf were closed, then E would be nearly zero. Let us find the leaf temperature for such a case, when

$$f(T) = Q_A - h_c(T - A) - \sigma \epsilon T^4 = 0, \tag{10.1}$$

$Q_A = 80$ mW cm^{-2}, $h_c = 4$ mW cm^{-2} deg^{-1}, $A = 300°$K, $\sigma = 5.67 \times 10^{-9}$ mW cm^{-2} deg^{-4}, and $\epsilon = 0.97$. Eqn. 10.1 tells us how badly our energy balance is out of balance, at various leaf temperatures. This means we seek the physically real root of the quartic polynomial plotted in Fig. 10.2, p. 232, i.e.,

$$f(T) = 5.5 \times 10^{-9} T^4 + 4T - 1280 = 0. \tag{10.2}$$

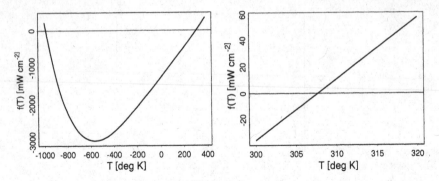

Figure 10.2: The quartic polynomial, Eqn. 10.2, that models the steady-state energy balance of a non-transpiring leaf (left), and another view (right) for temperatures just above the air temperature of 300° K.

All numerical methods for finding roots require one to choose a starting guess (or to guess a range) for the root. With the present problem, we can be reasonably confident that the leaf's temperature is near, but probably above, air temperature A, and so we could use that as a starting guess. However, theory tells us that a quartic function must have four roots in all, so here, as in most problems of finding roots, it proves worthwhile to plot the function against a range of T values to determine its general behavior before proceeding numerically as in Fig. 10.2 (left).

It is clear that, within the resolution of the figure, only one root lies anywhere near 300° K, and this knowledge makes finding the root more certain, as we shall see. Theory tells us further that if a quartic has one real root, then it must have at least two. Indeed, with the present function the graph shows another root near $T = -987.5°$, which is physically impossible, and certainly irrelevant for our purposes.

The other two roots of this equation are probably *complex conjugates* of the form $a \pm bi$, where $i = \sqrt{-1}$. These are also irrelevant. The other possibility is that the other two roots are real numbers, but lie outside the range plotted—this would also mean they were *physically meaningless*. In any case the root we seek is the one near 300° K. It is therefore useful to replot the function with greater resolution near this root, as in Fig. 10.2 (right). As seen there, our function is nearly linear for T between 300 and 320 degrees K, which will lead to fast convergence for the methods that follow, since most of them converge in a single iteration for truly linear functions.

Newton's Method

One common and generally efficient method for finding roots of non-linear equations is known as *Newton's method*. (An alternative name is the *Newton-Raphson* method.) This method can be derived easily from the theory of Taylor series. Recall that any nice function can be written as

$$f(x) = f(a) + f'(a)(x - a) + f''(a)(x - a)^2/2! + \dots.$$

from which $f(x) \approx f(a) + f'(a)(x - a)$ provided x is close enough to a. (Using such an approximation is called *linearization*.)

Now suppose that we want to find a root of some $f(x)$. We could

- choose a guess, x_0, and expand the function about $a = x_0$.

- truncate the resulting Taylor series after the linear term to get $f(x) \approx f(x_0) + f'(x_0)(x - x_0) = 0$.

- set this $f(x) = 0$ (because by definition the function will have that value at the root we seek) and solve the result for $x \approx x_0 - f(x_0)/f'(x_0)$. The result should be an improved estimate of the root, but it will be only approximate (except for a linear $f(x)$) because we have truncated the Taylor series.

- iterate to convergence (or to obvious divergence).

To apply this process to the leaf energy balance, we return to the form $f(T) = aT^4 + bT + c$, where $a = 5.5 \times 10^{-9}$, $b = 4$, and $c = -1280$. Note that the value of this $f(T)$ represents how much the net energy uptake of the leaf is out of balance, and so we wish to find the value of T that drives it to zero. Here $f'(T) = 4aT^3 + b$, so we iterate from our starting guess ($x_0 = 300$) using

$$T_{i+1} \approx T_i - \frac{aT_i^4 + bT_i + c}{4aT_i^3 + b} \quad \left[\text{of form } x_1 \approx x_0 - \frac{f(x_0)}{f'(x_0)}\right].$$

This process yields 300, 307.717, 307.678, 307.678. Of the various schemes we consider, Newton's method usually has the fastest rate of convergence. Its disadvantages are that it doesn't always converge if the initial guess is too far from the root, that you have to differentiate the function (without error!), and that it is not trivial to program in computers because of its use of the derivative.

Bisection

We turn now to a method with nearly opposite properties—*bisection* is simple and reliable, but slow and inefficient. Bisection differs from Newton's method in requiring *two* starting guesses (say x_0 and x_1), guesses that must *bracket* the root. This means that $f(x)$ must cross the x axis between x_0 and x_1, and so the product $f(x_0)f(x_1)$ must be negative[1]. Further, there must be only one root between x_0 and x_1, so to stay out of trouble it is best to plot $f(x)$ for $x_0 \leq x \leq x_1$.

Once we have two guesses that do bracket the root, we proceed as follows. We will illustrate the method with our leaf energy balance and with starting guesses of $T_0 = 300$, $T_1 = 320$. (You might wish to sketch lines on a copy of Fig. 10.2, right, to aid in understanding the method.)

- Bisect the T_0-T_1 range. I.e., calculate $T_2 = (T_0 + T_1)/2 = 310$. Then calculate $f(T_2)$ and choose the subrange, (T_0, T_2) or (T_2, T_1), that continues to bracket the root. Mathematically, the former will be true if $f(T_0)f(T_2) < 0$, as is the case here.

- Now bisect the new subrange, and calculate $f(305)$ in this instance. This gives us an even smaller subrange, (305–310) that we know contains the root.

- Continue bisecting, each time working with the latest interval that we know includes the root.

- Note that as we go, we pin down the root to within a range of width $|x_1 - x_0|$, $|x_1 - x_0|/2$, $|x_1 - x_0|/4$, $|x_1 - x_0|/8$, $|x_1 - x_0|/16$, and so on. In fact, after n bisections, we know that neither end of the current interval is farther than $|x_1 - x_0|/2^n$ away from the root.

---

[1] If you don't see why this is so, try sketching a function that crosses the $y = 0$ axis.

Table 10.1: Progress of root-finding with the bisection method applied to finding an equilibrium leaf temperature from Eqn. 10.2

T	f(T)	T	f(T)
300	-35.45	307.734	0.262459
320	57.67168	307.695	0.081169
310	10.79365	307.676	-0.00947
305	-12.4049	307.686	0.035849
307.5	-0.82514	307.681	0.013188
308.75	4.979344	307.678	0.001858
308.125	2.07588	307.677	-0.00381
307.813	0.625067	307.678	-0.00097
307.656	-0.10011	307.678	0.000442

In the present case, our calculations proceed as in Table 10.1. The iterations converge to a leaf-temperature estimate of 307.678° K. The $f(T)$ column indicates how well the energy flows are balanced at each temperature estimate. Although the values of T shown in the table are rounded to the nearest 0.001°, the calculations were actually carried out in a desktop computer with perhaps 16 or more digits of precision. Note the large number of iterations required to reach 0.001° precision with this method. Other methods, like Newton's above and the secant method below, are usually much more efficient, although they may not be as certain to locate a root.

The Secant Method

We close with one other method for seeking roots, a method that is intermediate in safety and efficiency, but easy to program. Again we begin with two guesses for the root, say x_1 and x_2. (It is safest, but not absolutely required, for these guesses to bracket the root.) We calculate $y_1 = f(x_1)$ and $y_2 = f(x_2)$, and seek x_3 such that $f(x_3) = 0$. Usually $f(x_3)$ will not *equal* zero, but we hope it will be closer to zero than either of its two predecessors.

The scheme we use is, in effect, to draw a straight line between the two starting points and to find the x intercept of that line (i.e., the point on that line where $y = 0$.) Of course, if $f(x)$ itself were a linear function, then this intercept would be the root of that function, and no further iterations would be needed. But since $f(x)$ is non-linear, the intercept will only be close (we hope) to the root we seek.

To find the intercept, we use the *two-point formula* for a straight line; i.e.,

$$\frac{y - y_1}{x - x_1} = \frac{y_2 - y_1}{x_2 - x_1} = \text{slope of line.}$$

If we now set $y = 0$, then $x_3 = x$ is the next approximation we seek. Thus

$$\frac{0 - y_1}{x_3 - x_1} = \frac{y_2 - y_1}{x_2 - x_1} = \text{slope of line,}$$

or

$$x_3 = x_1 - y_1 \left(\frac{x_2 - x_1}{y_2 - y_1} \right).$$

This is our general iteration formula, but we now need a way of deciding which of the previous two points to throw out. One scheme for making this decision results in the method of false position, which we will not consider further here. A second scheme yields the secant method. For the latter, one compares $|y_2|$ with $|y_1|$ and discards whichever point is farther from the x axis, i.e., the one with the larger absolute function value.

For the leaf energy balance, if we begin with $T_1 = 300°$ and $T_2 = 320°$, the secant method produces the following sequence[2]:

T_1	y_1	T_2	y_2	T_3	y_3
300	-35.45	320	57.6717	307.614	-0.2976
300	-35.45	307.614	-0.29759	307.678	0.00156
307.614	-0.2976	307.678	0.00152	307.678	-6.56E-08
307.678	0.00152	307.678	-6.56E-08	307.678	0

As predicted, this procedure has converged rapidly, in part because our $f(T)$ is so nearly linear between $T = 300°$ and $T = 320°$.

One question that arises with every iterative method is "when should we stop?" In the calculations just presented, the decision was easy, because $f(T)$ went to an actual zero (within the precision of the computer used), and no root can be better than that. Note that the

---

[2]Some values are shown at reduced precision so the table will fit the page.

T values in the table are rounded to the nearest 0.001 degree, but at full precision the values in the calculations continued to change until the last value, T_3, for which $f(T) = 0$.

Often, iterative methods for finding roots do not converge to the point where $f(x) = 0$ exactly, even after a large number of iterations. Rather, because of round-off error and limited machine precision, the iterations will lead to a cycle of x values for which $f(x)$ is only close to zero. On the other hand, frequently all that is needed is an estimate of the root that is good to a few digits of precision. If this is the case, then you can stop iterating as soon as $|x_{i+2} - x_{i+1}| < \epsilon$, where ϵ is the allowable error in the root. For the present problem, we might have decided to stop iterating once we knew the root to the nearest hundredth of a degree, say. In that case, we could have stopped after the third iteration.

There is another consideration, though. Sometimes, the precision of the root itself is less important than that the function value $f(x)$ be sufficiently close to zero. In the present problem, our criterion for stopping might have been that we wanted the net energy flows to be balanced to within 0.05 mW cm^{-2}. If so, we could have stopped after the second iteration, when $f(T) = 0.00152$. To carry this point a little further, in some problems one might want to set conditions on both the precision of the root estimate and on the absolute value of the corresponding function value. Double conditions like this are easily accommodated in subroutines that find roots by any of the methods we have studied.

Roots of Polynomials

Although we will not treat them in this book, be aware that special numerical methods are available for finding roots of polynomials. These are useful if your polynomial has repeated roots (see below), or if you need to find roots that are complex numbers. Most textbooks on numerical analysis (e.g., Press et al. 1992) discuss methods of this type.

Summary of Methods

To close our consideration of these various methods for finding roots of functions, we can summarize their relative efficiencies (converging with few function evaluations), reliabilities (likelihood of converging

Table 10.2: Comparison of properties of methods for finding roots of non-linear equations.

Method	Efficiency	Reliability	Simplicity
Bisection	low	high	high
Secant	medium	medium	medium
Newton	high	medium	low

to a solution), and simplicity of use (somewhat subjective) as in Table 10.2.

Finding Roots with MATLAB

To find the zero of the leaf energy balance (Eqn. 10.2) using MATLAB®, we could enter

```
enbal=inline('5.5e-9*T^4+4*T-1280');
Tsteady=fzero(enbal,300)
```

(The 300 in the call to fzero is optional, but providing a starting value helps MATLAB to find the root we seek.) The result would be returned as Tsteady = 3.076778222325283e+002, along with additional information about the convergence. The documentation for the fzero function describes other options available for using it.

Because the present function is a quadratic, we could also use MATLAB's "root" function to find its four roots. The call could take the form[3]

```
quartic=[5.5e-9 0 0 4 -1280];
roots(quartic)
```

The result of that call is

```
ans =
  1.0e+002 *
 -9.87493400340406
  3.39907789053938 + 8.064996801390019i
  3.39907789053938 - 8.064996801390019i
  3.07677822232528
```

---

[3]The "quartic" vector contains the coefficients of T^4, T^3, ..., T^0, in that order.

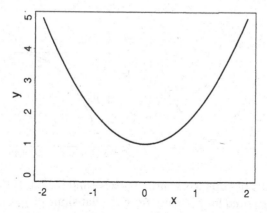

Figure 10.3: A quadratic, $y = x^2 + 1$, with no real roots.

As noted before, only the last of those roots is of physical interest in this case.

10.2 Repeated or Multiple Roots

Here I raise two cautions about finding roots. First, some equations may not have any real roots, as with $f(x) = x^2 + 1$ (Fig. 10.3), which equals zero only when $x = \pm\sqrt{-1}$. Thus its two roots are *complex conjugates*, not real numbers. The lack of real roots can be seen in the figure.

The second caution is that roots may sometimes be *repeated*. For example, $f(x) = x^3 - 3x^2 + 3x - 1$ can be factored into $f(x) = (x - 1)(x - 1)(x - 1)$, showing that the three roots of this cubic are +1, +1, and +1. That is, this root is repeated three times, making it very difficult to locate accurately—it is *ill-determined*. In a similar way, we can construct a cubic such as $y = f(x) = (x-1)(x-2)(x-2)$ with two different roots, one of which is repeated twice.

It is important to be aware of repeated roots because when they occur, they are often hard to find. Suppose for example, that

$$y = f(x) = (e^x - 5)(x - \log 5), \qquad . \tag{10.3}$$

as shown in Fig. 10.4 (left), p. 240. Given the function *stated in that form*, it is easy to see that it has two roots, as described in Eqns. 10.4 and 10.5.

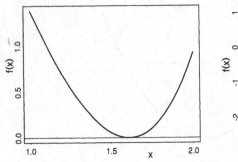 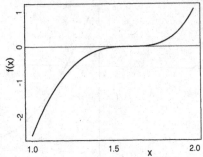

Figure 10.4: Plots of Eqns. 10.3 and 10.6 (left), and of Eqn. 10.7 (right), which has one root repeated in triplicate. Note the flat slope of $f(x)$ near the root.

1. $(e^x - 5) = 0 \Rightarrow e^x = 5 \Rightarrow x = \log 5$, and $\qquad$ (10.4)

2. $(x - \log 5) = 0 \Rightarrow x = \log 5.$ $\qquad$ (10.5)

Thus the root $x = \log 5$ is repeated twice.

If the function were presented to us in the form of Eqn. 10.3, we would have little trouble finding these roots and determining that they were the same. However, if $f(x)$ were multiplied out using the approximate numerical value of $\log 5 \approx 1.60944$, it might appear as:

$$f(x) = (x - 1.60944)e^x - 5x + 8.04719. \qquad (10.6)$$

Given this form of the function, I'm sure I would not recognize its factors (would you?), and I would have to use a numerical method to find its roots.

Note that if I tried to bracket the repeated root (1.60944) with $x_1 = 1$, $x_2 = 2$, I would obtain $f(x_1) = 1.39056$ and $f(x_2) = 0.93306$, both of which are positive. This problem shows up well if you just plot the function—are you beginning to get the idea that plotting a function can be an important step toward finding its roots?

Clearly, we can't use bisection, the secant method, or false position to find these roots; less clearly, Newton's method will often fail with repeated roots too.

One further example is illustrative. The function

$$f(x) = (e^x - 5)(x - \log 5)(3x - \log 125), \qquad (10.7)$$

shown in Fig. 10.4 (right), has the root $x = \log 5$ repeated *three* times. (Be sure you see why.) Here it *is* possible to bracket the root, but both

$f'(x)$ and $f''(x)$ are zero there, and the function is very flat at the root as well. Thus, even if bisection or the secant method would work in principle, functions like this give poor numerical results because of round-off difficulties. Finding repeated (multiple) roots is discussed in advanced numerical analysis texts. We will not cover it further, but be aware of the problem and *always plot your function!*

10.3 Exercises

1. Newton's method provides a handy way to obtain square roots iteratively. It is likely the method most calculators apply to get square roots, and you can use it effectively if you don't have a calculator that provides them. Show that Newton's method yields

$$x_{i+1} = \frac{x_i + N/x_i}{2} \tag{10.8}$$

   if you use it to find the root of $f(x) = x^2 - N$. Apply this method to find the square root of 17, starting with an estimate of 4.

   Note that Eqn. 10.8 calculates each new value as the arithmetic average of the previous guess and N over the previous guess. If one of those terms is too big, the other will be too small, and their mean comes closer to the value sought.

2. Many cooling processes can be modelled by "Newton's law of cooling", which states that, to a first approximation, the rate of heat loss from an object is proportional at any instant to the difference between its temperature and that of its surroundings. Mathematically, this can be described as:

$$\frac{d\theta}{dt} = -b\theta, \text{ from which } \theta(t) = \theta_0 e^{-bt},$$

   in which θ represents the temperature *difference* between an object and the air. Suppose two solar collectors are at the same temperature ($\theta_0 = 60°$ C) when a cloud bank rolls in at $t = 0$. The first collector has thermal properties that lead to a value of $b_1 = 1/3$ min^{-1}, while for the second, $b_2 = 1/2$ min^{-1}.

   How long will it take before the second collector is $2°$ C cooler than the first? Use the secant method. Use the secant method. To

demonstrate your ability to use the method correctly, your answer
should be correct to the nearest 0.001 minute.

3. Along similar lines, consider the temperature differences θ_1 and
   θ_2 for two collectors given by:

$$\theta_1(t) = \theta_{1,0}e^{-b_1 t} \text{ and } \theta_2(t) = \theta_{2,0}e^{-b_2 t}.$$

Suppose we want to determine how long it will take (from starting
time $t_0 = 0$) for $\theta_1(t)$ and $\theta_2(t)$ to differ by D Celsius degrees; thus
$D = \theta_1 - \theta_2$. For each of the following situations discuss whether,
and why, it would be better to obtain a numerical solution or an
analytic solution. (Also state which is *possible* in each case.)

a) In a particular case at hand, $\theta_{1,0} = \theta_{2,0} = 80°C$, $b_1 = 0.3$ min^{-1},
   $b_2 = 0.5$ min^{-1}, and $D = 2°$ C. However, we know we will be
   interested in other combinations of values in the future.

b) In a particular case at hand, $\theta_{1,0} = 75°C$, $\theta_{2,0} = 80°C$, $b_1 =$
   0.3 min^{-1}, $b_2 = 0.5$ min^{-1}, and we want to know how long it
   will take for the two collectors to cool to identical temperatures. Furthermore, we know that in all future cases we will
   only be interested in finding the time required for two collectors to cool until their temperatures are equal.

4. As you know, three-dimensional relationships of the form $z =$
   $f(x,y)$ can be plotted as a contour diagram in two dimensions;
   i.e., as a "contour map" of "elevations" z, drawn in the x-y plane.
   Computer programs to make such plots for arbitrary $z = f(x,y)$
   functions can be written as in the following example:

Suppose $z = f(x,y) = 0.7(xy)^2 - 4(xy)^{0.5}$. For this function,
you can plot a given contour (for $z = 3$, say) by:

a) setting $z = 3$, then

b) choosing a particular y value ($y = 2$, say), and

c) calculating the corresponding x for which $3 = f(x,2)$. Then
   you repeat Step C for many y values, and finally, repeat the
   whole procedure for each z contour that you want to plot.

To aid in understanding this procedure, calculate x such that
$f(x,2) = 3$. Use a numerical method. Note that this gives you

just one point on the $z = 3$ contour (above $x = 2$); the computer would have to repeat the calculation many times to allow plotting the contour.

5. The oxygen deficit D in a stream caused by a wastewater discharge is sometimes modelled using the solution to the Streeter-Phelps equations, which you have seen before. One form is:

$$D(t) = \frac{kB_0}{r - k}(e^{-kt} - e^{-rt}) + D_0e^{-rt}$$

where D = oxygen deficit (relative to saturation) [mg L^{-1}], D_0 = initial value of D at the point of discharge, B_0 is the initial value of BOD concentration at the point of discharge [mg L^{-1}], k is an oxidation rate coefficient [da^{-1}], and r is a re-aeration coefficient [da^{-1}].

Suppose that for a particular discharge

$B_0 = 6.75$ mg L^{-1}
$D_0 = 1.59$ mg L^{-1}
$k = 0.607$ da^{-1}
$r = 0.76$ da^{-1}.

a) Plot $D(t)$ for t between 0 and 3 days.

b) Suppose further that Pumpkinseed Sunfish cannot survive in the stream if the oxygen deficit $D > 2$ mg L^{-1}. Also suppose the stream flows at a velocity of 1.5 ft s^{-1}. Then determine the *range of distances* downstream from the discharge point where these fish could not survive.

6. A system of two equations for the transfers of a lipid-soluble chemical between a person's blood and fatty tissues, when there is a constant rate of intake of the chemical, might take the form:

$$\frac{dC_b}{dt} = \frac{U}{B} - fC_b - \frac{k}{B}\left(C_b - pC_f\right) \text{ and } \frac{dC_f}{dt} = \frac{k}{F}\left(C_b - pC_f\right),$$

where C_b and C_f are the concentrations in the blood and fat, respectively, U is the intake rate, B and F are the masses of blood and fat in the body, p is a partition coefficient accounting for differential solubility of the substance in the two tissue types, and f and k are rate constants.

If we set the uptake rate to zero, we could apply that model to the decline of dioxin in the body of the Ukrainian politician, Viktor Yushchenko, who was reportedly poisoned with that substance in 2004. As rough guesses, let's set $B = 10$ kg, $F = 15$ kg, $k = 0.1$ kg da^{-1}, $f = 0.01$ da^{-1}, $C_b(0) = 1$ mg kg^{-1} (i.e., 1 ppm), $C_f(0) = 10$ mg kg^{-1}, and $p = 0.1$ [unitless], at some starting time. MATLAB's dsolve command yielded this approximate solution for his blood concentration (in mg kg^{-1}) as it might have varied after time zero:

$$C_b(t) \approx 0.733 \exp(-0.00618t) + 0.267 \exp(-0.0205t).$$

Using Newton's or the secant method[4], estimate how long it would have taken for the concentration of dioxin in Yushchenko's blood to decline from its estimated starting value of 1.0 mg kg^{-1} to 0.1 mg kg^{-1}. Provide the result to the nearest 0.01 day for checking purposes, even though that is excessive accuracy for the real problem.

You will likely find this plot useful:

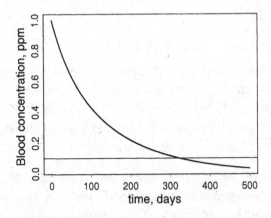

7. The utility department of a large city with several reservoirs in its water supply decides to use copper sulfate in two of them one summer, to kill off some aquatic weeds. The weed species in the two water bodies are different, so different maximum concentrations are needed for the two reservoirs. In particular, Reservoir A will require a concentration of 80 mg L^{-1} and Reservoir B will

---

[4]Be sure to review the bold-faced question on p. 230.

require a concentration of 60 mg L^{-1}. The $CuSO_4$ will be applied around midday on the same day in both reservoirs.

The two water bodies have the following flow characteristics:

Res.	Volume $[m^3]$	Flow through $[m^3 \, da^{-1}]$
A	55,000	2,700
B	38,000	700

(Assume that both reservoirs are well mixed, and that the volumes and flows for each remain constant through time.)

Your supervisor asks you to determine how long it will be until the Reservoir A concentration is 1 mg L^{-1} below that of Reservoir B. That is your task here. Be sure to include at least one unit check for any analysis you perform.

Hint: You will need to apply ideas from more than one chapter of the book to this question. Specifically, the $CuSO_4$ concentration in each reservoir should drop exponentially from its initial concentration, with an appropriate "decay rate" for each water body. You'll also need a root-finding method (you may choose one) to determine when the two curves differ by the stated amount.

8. The MPN (most probable number) method is sometimes used to estimate concentrations of microorganisms in situations where the organisms are difficult to count as individuals, but where it is feasible to determine their presence or absence in a sample. For example, the method can be used to estimate concentrations of methane-consuming bacteria (methanotrophs) in soils, by analyzing for methane disappearance from sample tubes.

Using the MPN method involves repeating serial dilutions of the medium under study, and determining the number of samples at each dilution level in which the quality of interest is detected. Often the results are determined from tables that exist for the purpose, but in some cases, you may need to use a combination of sample number or dilution factor for which you can't find tables. In such a case, it is possible to determine the most probable number of organisms present in samples of the lowest dilution level using a formula like the one here. (This particular version applies to a special case with only two dilution levels, to minimize the amount of computation you need to do. Usually more levels are involved.)

The formula is:

$$\frac{a_1 p_1}{1 - e^{-a_1 x}} + \frac{a_2 p_2}{1 - e^{-a_2 x}} = a_1 n_1 + a_2 n_2, \tag{10.9}$$

where a_i is the ith dilution level, p_i is the number of *positive* tubes at dilution i, i.e., the number exhibiting the response of interest, x is the most probable number (MPN) of organisms in the level 1 samples, and n_i is the number of tubes tested at dilution level i.

A microbiologist in the lab where you work asks you to determine the MPN (i.e., x) of methanotrophs at the 1/10 dilution, for a study in which $n_1 = n_2 = 5$, $a_1 = 0.1$, $a_2 = 0.01$, $p_1 = 4$, and $p_2 = 1$. Determine this value to the nearest 3 significant digits, showing your work in detail. You may use any appropriate method, but identify the method. A plot of the left hand side (LHS) of the equation above is shown in Fig. 10.5.

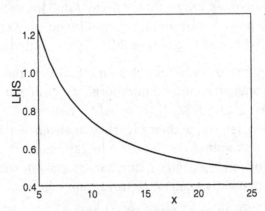

Figure 10.5: Plot of the LHS of Eqn. 10.9

9. Spruce Lake, which feeds into Pine Lake, becomes contaminated with a substance from a one-time spill. State hydrologists derive differential equations that describe the tranfer from Spruce into Pine and the subsequent flushing of Pine, and estimate that the concentration of the substance [mg L^{-1}] in Pine should vary roughly as

$$C(t) = 120 \left[e^{-0.27t} - e^{-0.4t} \right].$$

The rate constants in the exponentials have units of yr^{-1}.

The state Department of Environmental Protection wishes to warn residents on Pine Lake not to use the water from that lake when the concentration of the contaminant rises above 15 mg L^{-1}. Use any of the methods described in this chapter to estimate when that would first occur. You'll likely find it helpful to plot the concentration over the range 0–10 yr.

10.4 Questions and Answers

1. How can you be sure you are bracketing the root for these methods?

   • Note that the bracketing concept is relevant only for the methods that require you to have two starting guesses. It doesn't apply with Newton's method.

   If you have two guesses, x_1 and x_2, and if you are trying to find the zero(s) of $f(x)$, then you know you are bracketing a root (or three, or five, or some odd number of roots) if one of $f(x_1)$ and $f(x_2)$ has a plus sign and the other has a minus sign. Draw a sketch of an arbitrary $f(x)$, and convince yourself of that. This condition is equivalent to the product $f(x_1) \times f(x_2)$ being negative.

2. Why do we want to know the roots of polynomials?

   • The leaf energy balance (which could be the energy balance of just about anything, like the top of your car, or ...) was a polynomial, the root of which is the leaf temperature. If you were studying carbon balances of forests, you might want to know leaf temperatures because they would affect photosynthesis rates (for example). So, in modelling, you'd find the temperature by finding the root of the energy balance equation. Polynomials often come up in many other applications, such as risk analysis, cost-benefit analysis, and so on.

3. Why does the secant method have that name?

   • I don't know. The straight line we fit between the latest two points probably has something to do with a secant, but I'm not sure what the relationship is.

4. Comment: It seems that Taylor series are among the most applicable tools we've learned this semester.

   • I agree. They form the basis for much of mathematical analysis, even when they aren't mentioned explicitly.

5. When you used the bisection method, what function were you using for $f(T)$?

   • Since you've written T, it would have been the leaf energy balance in the form $f(T) = 0$. In general, you would apply bisection to find the value of x that makes some $f(x)$ equal to *zero*.

6. Please explain again, graphically, how the secant method works. Which value of the x variable do you throw out at each step?

   • Sketch some curve that starts out at low x, where $f(x)$ is negative, and curves up to high x, where $f(x)$ is positive. You then pick one point (x_1,y_1, say) to the left of where the curve crosses the x axis, and a second point (x_2,y_2) to the right of that crossing. Then draw a straight line between the two points. Where that line crosses the x axis is the x_3 value, and y_3 is the value of the function at that value of x_3. With the secant method, at each iteration, after you calculate x_3 and $y_3 = f(x_3)$, you replace whichever of x_1 or x_2 had a function value with the largest absolute value.

7. Do you need to draw the graph of a non-linear function before finding its root(s)?

   • That isn't mandatory, but I *strongly* recommend it. Otherwise you may not know whether there is more than one root that might be of interest. Graphs can also help you to choose good starting values and to detect problems like multiple roots.

8. What kind of accuracy is achievable with Newton's method?

   • The accuracy of all these methods is limited primarily by the number of digits stored for a variable in the computer or calculator you are using. You can iterate as many times as needed to improve accuracy, but eventually round-off error will prevent doing any better. In Maple, you can set Digits=50, say, to get greater accuracy if you need it, but I don't think MATLAB has the same ability. You can presumably get better answers from a spreadsheet than with an 8-digit calculator. In programming languages like Fortran, you can do better if you use double-precision (or even quad-precision) variables in place of single-precision variables.

9. Comment: In addition to programs like MATLAB, Maple, Mathematica and the like, spreadsheets now contain "solve" functions as well. In Quattro, it's in Tools|Numeric Tools|Solve For. In Excel, try Tools|Solver.

Chapter 11

Partial Differential Equations

11.1 Partial Derivatives

We begin this chapter with an introduction to the meaning of partial derivatives, and a discussion of the variety of notations used to represent them. Because we will use a new symbol (∂)[1], these derivatives may look complicated, but they aren't really very different, in the ways we'll use them, from the ordinary ones you already know about.

Consider a model for the growth rate G of algal biomass in a pond or lake, in units like mg L^{-1} day^{-1}. In this model, the growth rate increases as the product of a logistic in nitrogen concentration N [mg L^{-1}] times a second logistic in phosphorus concentration P. Such a model might be of the form:

$$G(N,P) = \left(\frac{K_n}{1 + ae^{-r_nN}} + b \right) \cdot \left(\frac{K_p}{1 + ce^{-r_pP}} + d \right), \tag{11.1}$$

which is plotted in Fig. 11.1, and as a contour plot in Fig. 11.2 (p. 251 for both).

A limnologist might want know how much this algal growth rate would increase if N or P increased by small amounts. These quantities, which would depend on the current values of N and P, could handily be expressed as:

---

[1] This symbol is usually read as "partial," or sometimes just as "d."

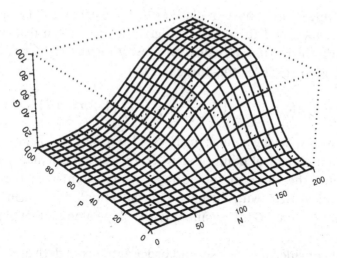

Figure 11.1: A hypothetical relationship (Eqn. 11.1) between algal growth rate G and nitrogen and phosphorus concentrations in a pond or lake. See also Fig. 11.2.

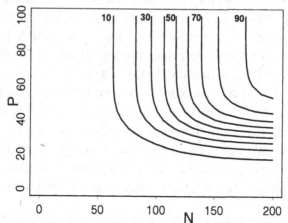

Figure 11.2: Contour diagram of the relationship plotted in Fig. 11.1. The contour values of 10-90 are values of growth rate.

Increase in growth rate per increase in $N = \dfrac{dG}{dN}$, and

Increase in growth rate per increase in $P = \dfrac{dG}{dP}$.

Actually, these expressions come close to what we want, but because G varies with both N and P, it is conventional to use the "∂"

symbol referred to above in place of d. If we act as if P is temporarily a constant, and then take the derivative of G with respect to N, we obtain the *partial derivative of G with respect to N*; i.e., $\partial G/\partial N$. According to MATLAB, this turns out to be

$$\frac{\partial G}{\partial N} = aK_n r_n e^{-r_n N}\left(\frac{K_p}{1 + ce^{-r_p P}} + d\right)\left(1 + ae^{-r_n N}\right)^{-2};\qquad (11.2)$$

$\partial G/\partial P$ would have a similar form. (See Exercise 1, p. 272).

From a biological point of view, $\partial G/\partial N$ at a particular N-P point represents the slope of the surface at that point (i.e., the rate of change of growth with nitrogen concentration), along a line parallel to the N axis; $\partial G/\partial P$ is similarly the slope along a line parallel to the P axis.

Next consider the four second-order and mixed derivatives:

$$\frac{\partial^2 G}{\partial N^2},\ \frac{\partial^2 G}{\partial P^2},\ \frac{\partial^2 G}{\partial N \partial P} = \frac{\partial}{\partial N}\left(\frac{\partial G}{\partial P}\right),\ \text{and}\ \frac{\partial^2 G}{\partial P \partial N} = \frac{\partial}{\partial P}\left(\frac{\partial G}{\partial N}\right).$$

The first of those describes the rate at which $\partial G/\partial N$ increases as one moves along a constant P line (parallel to the N axis), and the second the rate at which $\partial G/\partial P$ increases along a constant N line (parallel to the P axis). The third describes the rate at which $\partial G/\partial P$ increases along a constant P line (parallel to the N axis), and the fourth the rate at which $\partial G/\partial N$ increases along a constant N line (parallel to the P axis). Try moving an imaginary ruler around on the Fig. 11.1 surface in your mind, to get a better feeling for those meanings.

Be aware that other notation is often used for partial derivatives. For example, alternative symbols for derivatives like those just given, for a function $z = f(x, y)$, are (Edwards and Penney 1982):

$$\frac{\partial z}{\partial x} = \frac{\partial f}{\partial x} = f_x(x, y) = f_1(x, y) = D_x f(x, y) = D_1 f(x, y);$$

$$\frac{\partial z}{\partial y} = \frac{\partial f}{\partial y} = f_y(x, y) = f_2(x, y) = D_y f(x, y) = D_2 f(x, y);$$

$$(f_x)_x = f_{xx} = \frac{\partial f_x}{\partial x} = \frac{\partial}{\partial x}\left(\frac{\partial f}{\partial x}\right) = \frac{\partial^2 f}{\partial x^2} = \frac{\partial^2 z}{\partial x^2};$$

$$(f_x)_y = f_{xy} = \frac{\partial f_x}{\partial y} = \frac{\partial}{\partial y}\left(\frac{\partial f}{\partial x}\right) = \frac{\partial^2 f}{\partial y \partial x};\ \text{and so on.}$$

Another symbol is sometimes used for certain combinations of partial derivatives, namely the ∇ *operator*, which is usually referred to as *del* in the U.S., and as *nabla* in Europe. This operator is especially useful for various vector relationships (e.g., Schey 1973). An example of its use in a non-vector context is

$$\nabla^2 T = \frac{\partial^2 T}{\partial x^2} + \frac{\partial^2 T}{\partial y^2} + \frac{\partial^2 T}{\partial z^2}.$$

This is known as *del-squared* of T (temperature) or the *Laplacian* of T, and is useful shorthand for a combination of derivatives that occurs frequently in applied math. For example, it helps describe heat transfer by conduction in a solid, as we shall see. If T is replaced by the concentration C of some substance, then $\nabla^2 C$ appears in models of diffusion.

We close this section with some straightforward examples of partial derivatives. For the function

$$z = f(x, y) = 3x^2 - 2xy + y^5 :$$

$$\frac{\partial z}{\partial x} = 6x - 2y, \quad \frac{\partial z}{\partial y} = -2x + 5y^4,$$

$$\frac{\partial^2 z}{\partial x^2} = \frac{\partial}{\partial x}\left(\frac{\partial z}{\partial x}\right) = \frac{\partial}{\partial x}(6x - 2y) = 6,$$

$$\frac{\partial^2 z}{\partial y^2} = \frac{\partial}{\partial y}\left(\frac{\partial z}{\partial y}\right) = \frac{\partial}{\partial y}\left(-2x + 5y^4\right) = 20y^3,$$

$$\frac{\partial^2 z}{\partial x \partial y} = \frac{\partial}{\partial x}\left(\frac{\partial z}{\partial y}\right) = \frac{\partial}{\partial x}\left(-2x + 5y^4\right) = -2, \text{ and}$$

$$\frac{\partial^2 z}{\partial y \partial x} = \frac{\partial}{\partial y}\left(\frac{\partial z}{\partial x}\right) = \frac{\partial}{\partial y}(6x - 2y) = -2.$$

Note the equality of the two mixed derivatives. Calculus texts (e.g., Edwards and Penney 1982) often present proofs that the mixed derivatives are independent of order of differentiation for functions having derivatives that are suitably continuous. (In the rest of this book, we will not be using mixed derivatives anyway.)

11.2 Mass and Heat Transfer

In Chapters 4–8 we dealt with one or more responses that varied only with time at a given place, or along one dimension at a given

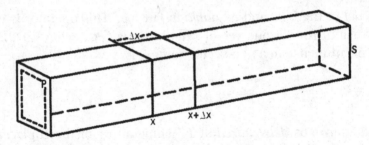

Figure 11.3: Hollow diffusion tube showing "control volume, of length Δx. Each side is of length S cm, and the perimeter is $P = 4S$. Alternatively, the figure can be viewed as a solid metal rod carrying an electrical current, for derivation of the transient (time-varying) heat conduction equation.

time. However, many important environmental variables like temperatures or pollutant concentrations may change simultaneously with time and along one or more spatial dimensions. In these cases we need to use *partial* differential equations (PDEs), because they allow us to deal with variations of the response variables in relation to two or more "independent" variables.

We will consider PDEs for one general class of problems, those involving transfers of mass, heat, or organisms through air, water, and solids. Because PDEs are usually much more complicated to solve than are ODEs, we will only touch on solution methods in this book. Our purposes will be to help you understand where PDEs come from, and what they mean. This should aid you in reading literature that includes PDEs as mathematical models for environmental phenomena. Many PDEs are nothing more than mathematical statements of mass balances or energy balances, and knowledge of that fact should aid your understanding.

For an example of a mass-balance situation leading to a PDE, we will generalize the diffusion problem from Chapter 8 to include time dependence. For the derivation, refer to Fig. 11.3.

We retain all features of the problem treated in Chapter 8, including adsorption of chloroform onto the tube wall, with one exception— now we suppose that $C(0) = C_1$ is not a constant but a specified function of time, $C_1(t) = f(t)$. We will still hold $C(L)$ constant at a fixed value C_2 to keep matters simple, but in principle it could vary with time as well. This one change means that everywhere in the tube (except at $x = L$), the local chloroform concentration will vary with both x and t. Thus, in mathematical notation, $C = C(x, t)$. As a result, the mass of chloroform in the control volume (CV), the diffusion rates into and out of that volume, and the rate of adsorption onto its walls, all vary with location and time as well. We can account for these changes by writing, for some short period of time Δt,

mass present in CV at time $t + \Delta t$ =
mass present in CV at time t
+ mass entering at x during Δt period
− mass leaving at $x + \Delta x$ during Δt period
− mass adsorbed onto CV wall during Δt period.

In symbols, this mass balance for the CV (located at x) becomes something like:

$$m(t + \Delta t) \approx m(t) + \left[-DA_{xs} \frac{dC}{dx} \bigg)_x \right] \Delta t$$

$$- \left[-DA_{xs} \frac{dC}{dx} \bigg)_{x+\Delta x} \right] \Delta t - 4kS\Delta x C \Delta t.$$

A unit check yields

$$mg \approx mg + \left(\frac{cm^2}{s} \right) cm^2 \left(\frac{mg}{cm^4} \right) s$$

$$- \left(\frac{cm^2}{s} \right) cm^2 \left(\frac{mg}{cm^4} \right) s - \left(\frac{cm}{s} \right) cm\ cm \left(\frac{mg}{cm^3} \right) s.$$

Cancellation shows that all terms have units of mg.

In Chapter 8, concentration gradients along a single dimension led to ordinary derivatives. Now, because C varies with both x and t— i.e., $C = C(x, t)$—the concentration gradients in this equation should be expressed as *partial* derivatives, $\partial C / \partial x$, since the instantaneous diffusion rates depend on the rate of variation of C with x and not its variation with t. Don't puzzle over this too much—it's just a matter of definition. In general, the partial derivative of $f(x, y)$ with respect to x is just the derivative of f with respect to x, treating y temporarily as a constant. Using partial-derivative notation, then, our equation becomes:

$$m(t + \Delta t) \approx m(t) + \left[-DA_{xs} \frac{\partial C}{\partial x} \Big)_x \right] \Delta t$$

$$- \left[-DA_{xs} \frac{\partial C}{\partial x} \Big)_{x+\Delta x} \right] \Delta t - 4kS\Delta x C \Delta t.$$

If we move the $m(t)$ term to the left, then the LHS refers to mass, and the RHS refers to concentration. To be more consistent, we can divide through by the volume of the CV, namely $A_{xs}\Delta x$, and rearrange some of the terms to obtain

$$C(t + \Delta t) - C(t) \approx \left[D \frac{\frac{\partial C}{\partial x}\Big)_{x+\Delta x} - \frac{\partial C}{\partial x}\Big)_x}{\Delta x} - \frac{4kS\Delta x C}{A_{xs}\Delta x} \right] \Delta t.$$

As always, we should check the units of the terms in this equation:

$$\frac{mg}{cm^3} = \left[\frac{cm^2}{s} \frac{mg}{cm^5} - \frac{cm}{s} cm \frac{mg}{cm^3} \frac{1}{cm^2} \right] s.$$

Now recall that $A_{xs} = S^2$, divide through by Δt, and cancel Δx from the last term to obtain:

$$\frac{C(t + \Delta t) - C(t)}{\Delta t} \approx D \frac{\frac{\partial C}{\partial x}\Big)_{x+\Delta x} - \frac{\partial C}{\partial x}\Big)_x}{\Delta x} - \frac{4k}{S} C.$$

Finally, if we take limits as both $\Delta t \to 0$ and $\Delta x \to 0$, we have

$$\frac{\partial C}{\partial t} = D \frac{\partial^2 C}{\partial x^2} - \frac{4k}{S} C.$$

This *partial differential equation* (PDE) represents the changes in the mass balance of the diffusing gas in the tube as the concentration varies with time and with x. Similar equations, based on an analogy between dispersion of organisms and diffusion of molecules, are used as mathematical models by population biologists. For an introduction to this topic, see Holmes, Lewis, Banks, and Veit (1994).

We won't attempt to solve this equation, at least not yet, but to do so, you would need to have one initial condition (IC), because there's a first derivative with respect to time; and two boundary conditions (BCs), because there's a second derivative with respect to distance. For example, the IC might take the form $C(x, 0) = f_i(x)$ for $0 \leq x \leq L$, which gives the concentration at every value of x at time zero. The BCs might take the form $C_1 = f_1(t)$ and $C_2 = f_2(t)$ at $x = 0$ and $x = L$ respectively. Some authors would use different terminology, calling these conditions $C(0, t)$ and $C(L, t)$. These particular BCs state the concentrations at the end points for all future times.

Heat Transfer by Conduction

Heat transfer by the process of conduction is physically analogous to molecular diffusion, and, not surprisingly, the mathematics of the two processes is similar as well. Now we derive a PDE to describe conductive heat transfer along a solid rod, as diagrammed by viewing Fig. 11.3, p. 254 in that way.

In particular, we consider a case when an electrical current along the rod produces a uniform (but possibly time dependent) rate of heat generation $G(t)$ per unit volume of rod material, where G has units of J cm^{-3} s^{-1}. Similarly to diffusion of a gas across a plane (p. 168), heat conduction across a plane follows the relationship:

$$Q = -kA\frac{dT}{dx}$$

where

$Q = $ heat flux [J s^{-1}]

$k = $ thermal conductivity [J cm^{-1} s^{-1} C^{-1}]

$A = $ area of the plane [cm^2]

$T = $ local temperature [C]

$x = $ distance along the rod [cm].

For BCs, we specify, for $T(x, t)$, that $T(0, t) = f_1(t)$ and $T(L, t) = f_2(t)$; i.e., that the temperatures at the two ends may vary over time as described by the functions f_1 and f_2. Then, for conduction along a rod with square cross-section (side S), a heat balance for a control volume between x and $x + \Delta x$ can be stated as:

heat in the CV at time $t + \Delta t = $
heat in the CV at time t
+ heat entering the left face during the Δt period

+ heat generated in the CV during the Δt period

– heat leaving the right face during the Δt period.

Mathematically, this balance becomes:

$$H(t + \Delta t) \approx H(t) + \left[-kA\frac{\partial T}{\partial x}\right)_x\right]\Delta t$$

$$- \left[-kA\frac{\partial T}{\partial x}\right)_{x+\Delta x}\right]\Delta t + GA\Delta x\Delta t,$$

where H is the heat content of the CV, in Joules. A unit check yields

$$J = J + \left[\frac{J}{cm\ s\ °C}\ cm^2\ \frac{°C}{cm}\right]s$$

$$- \left[\frac{J}{cm\ s\ °C}\ cm^2\ \frac{°C}{cm}\right]s - \left[\frac{J}{cm^3\ s}\ cm^2\ cm\ s\right].$$

From this,

$$\frac{H(t + \Delta t) - H(t)}{\Delta t} = \frac{\Delta H}{\Delta t} \approx kA\left[\frac{\partial T}{\partial x}\right)_{x+\Delta x} - \frac{\partial T}{\partial x}\right)_x\right] + GA\Delta x.$$

When we worked with diffusion of a gas, we made use of the relationship between mass m and concentration C in some volume V,
but that was easy because $C = m/V$, or $m/C = V$. Now we need a
similar relationship between heat content H (analogous with mass)
and "heat concentration," which is related to the temperature T. The
necessary relationship comes from physics, and turns out to be

$$\frac{dH}{dT}\left(\approx \frac{\Delta H}{\Delta T}\right) \overset{\text{def}}{=} c_p\rho V \text{ or } \frac{dT}{dH}\left(\approx \frac{\Delta T}{\Delta H}\right) = \frac{1}{c_p\rho V},$$

where c_p is the heat capacity (or specific heat) of the conductive
medium[2] [J g^{-1} C^{-1}] and ρ is the density of the medium [g cm^{-3}].

If the medium stays within a limited temperature range, so that
$c_p\rho V$ stays nearly constant, then

$$\Delta H \approx c_p\rho V(\Delta T).$$

Thus

---

[2]Note that "medium" is the singular form, and "media" is plural—many people
use these terms incorrectly.

$$H(t + \Delta t) - H(t) \approx c_p \rho V [T(t + \Delta t) - T(t)],$$

so

$$c_p \rho V \left[\frac{T(t + \Delta t) - T(t)}{\Delta t} \right] \approx kA \left[\frac{\partial T}{\partial x} \Big)_{x+\Delta x} - \frac{\partial T}{\partial x} \Big)_x \right] + GA\Delta x.$$

Now $V = A\Delta x$, so

$$\frac{T(t + \Delta t) - T(t)}{\Delta t} \approx \frac{k}{c_p \rho} \left[\frac{\frac{\partial T}{\partial x} \Big)_{x+\Delta x} - \frac{\partial T}{\partial x} \Big)_x}{\Delta x} \right] + \frac{G}{c_p \rho},$$

which, in the limit as $\Delta t \to 0$ and $\Delta x \to 0$, becomes

$$\frac{\partial T}{\partial t} = \frac{k}{c_p \rho} \frac{\partial^2 T}{\partial x^2} + \frac{G}{c_p \rho}.$$

Solution of this equation, for conduction in the presence of a heat source, would require an initial condition, $T(x, 0)$, in addition to the two BCs already specified. ●

11.3 Schmidt's Graphical Method for Solving PDEs

Although solving partial differential equations is in general beyond our scope, certain equations in which the main quantity of interest (e.g., temperature) varies with time and along a single physical dimension can be solved approximately by a simple graphical method. Because this solution process provides insight into the meaning of PDEs and their solutions, we will take up an example:

A biologist studying a rare species of shrew wants to model how temperatures change with time at various depths in the soil where the shrew lives. She uses a probe to measure the temperature profile once at a starting time, but wishes to avoid that disturbance later. She thus hopes to estimate sub-soil temperatures for other times, based on measurements of the surface temperature only.

To do this, we need the partial differential equation that describes time-varying heat conduction in one physical dimension. That equation (when there are no heat sources or sinks in the soil) is

$$k \frac{\partial^2 T}{\partial z^2} = c\rho \frac{\partial T}{\partial t} \quad \text{or} \quad \frac{\partial^2 T}{\partial z^2} = \frac{1}{a} \frac{\partial T}{\partial t}. \tag{11.3}$$

This is like the equation for conduction in the wire, but here there is no internal source of heat. The symbols here are:

$T = T(z, t)$ = soil temperature [C] at depth z [m] and time t [da],

k = thermal conductivity of the soil [J m^{-1} da^{-1} C^{-1}],

c = specific heat capacity of the soil matrix [J kg^{-1} C^{-1}],

ρ = density of the soil matrix [kg m^{-3}], and

$a = k/(c\rho)$ = *thermal diffusivity* of the soil [m^{-2} da^{-1}].

For some problems with fairly simple initial and boundary conditions, Eqn. 11.3 can be solved analytically, in the form of infinite series of sines and cosines that are known as Fourier series. The theory behind those solutions is outside our scope, so we will have a look at a relatively simple numerical-graphical technique instead.

Before going further, we need to take up a method for approximating a second derivative (ordinary or partial) as a finite difference. Suppose, for example, that we want to approximate

$$\frac{\partial^2 T}{\partial x^2} = \frac{\partial}{\partial x}\frac{\partial T}{\partial x}$$

at a point x_1, using a difference $\Delta x = h$. A common way to do this is to apply the forward difference scheme (p. 33) three times, as follows:

$$\frac{\partial^2 T}{\partial x^2} \approx \frac{\left.\frac{\partial T}{\partial x}\right)_{x+h} - \left.\frac{\partial T}{\partial x}\right)_x}{h}$$

$$\approx \frac{\frac{T(x+h) - T(x)}{h} - \frac{T(x) - T(x-h)}{h}}{h} \tag{11.4}$$

$$= \frac{T(x+h) + T(x-h) - 2T(x)}{h^2}.$$

Note that Eqn. 11.4 helps to show why $\partial^2 T/\partial x^2$ has units of deg cm^{-2}.

Equations like 11.3 can be represented approximately by using finite approximations, and considering changes in temperature over discrete steps of z and discrete steps of time. We will use subscripts (e.g., $i = 1, 2, \ldots$) to indicate successive values of depth z, and "overscripts" to represent values of time. In the discrete form, as we go from time t to time $t + \Delta t$ at physical location i, Eqn. (11.3) becomes

$$\frac{\overset{t}{T_{i-1}} + \overset{t}{T_{i+1}} - 2\overset{t}{T_i}}{(\Delta z)^2} \approx \frac{1}{a} \frac{\overset{t+\Delta t}{T_i} - \overset{t}{T_i}}{\Delta t}. \tag{11.5}$$

As you can see, the temperature variation through time is approximated by the RHS essentially as an Euler method step.

If we multiply through by $a\Delta t$, add $\overset{t}{T_i}$ to both sides, and solve for $\overset{t+\Delta t}{T_i}$, Eqn. 11.5 becomes

$$\overset{t+\Delta t}{T_i} \approx \overset{t}{T_i} + \frac{a\Delta t}{(\Delta z)^2}\left[\overset{t}{T_{i-1}} + \overset{t}{T_{i+1}} - 2\overset{t}{T_i}\right]. \tag{11.6}$$

This expresses the new temperature (at time $t + \Delta t$) at location i in terms of the old temperature (at time t) at that location and the old temperatures on either side of it. A physicist named Schmidt noticed that certain particular choices for Δz and Δt could simplify calculations involving that equation considerably. In particular, we can choose these two differences to force

$$\frac{a\Delta t}{(\Delta z)^2} = \frac{1}{2}, \text{ from which } \Delta t = \frac{(\Delta z)^2}{2a},$$

and since it is usually convenient to specify Δz based on the geometry of the system, this amounts to determining a value for Δt after choosing Δz. With the Δt set in that way, Eqn. 11.6 becomes

$$\overset{t+\Delta t}{T_i} \approx \frac{\overset{t}{T_{i-1}} + \overset{t}{T_{i+1}}}{2},$$

i.e., *the new temperature at location i is just the average of the old temperatures on either side of that location!* This would make calculations simple (even in a spreadsheet), but it also allows for a simple graphical solution of heat transfer problems like ours, and of related diffusion and population-migration problems.

Here is an example of Schmidt's graphical method. To interpret it quantitatively, we will use $a = 0.42$ cm^2 min^{-1}, and $\Delta z = 5$ cm, say[3]. Then each time step we take will represent

$$\Delta t = \frac{(\Delta z)^2}{2a} = \frac{5^2}{2(0.42)} = 0.496 \text{ hr} \approx 0.5 \text{ hr, or } 30 \text{ min.}$$

[3] The value for a is taken for loess soil from Geiger (1966), who indicated that this parameter varies widely with soil type. In a real situation, one would want a good estimate for the particular soil type present at the site under investigation.

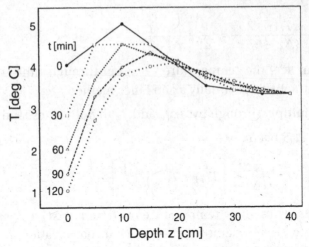

Figure 11.4: Schmidt's method applied to varying temperatures in a soil profile. The nine measured temperatures at time 0 (solid line) define the initial condition; and the five measured temperatures at depth 0, and the assumed constant value at 40 cm define the boundary conditions.

Suppose the initial temperature profile measured by the probe is as shown by the solid line in Fig. 11.4, and that the measured surface temperatures at several later times (t_1, t_2, ...) are as indicated along the left (temperature) axis. Then the temperature profiles estimated by Schmidt's method are indicated by the various dashed lines in the figure. We are making one major assumption here, that 50 cm is deep enough in the soil that the temperature there changes little over the time period of interest.

Note that as the soil cools at the surface, the soil just beneath it cools too, but as you go deeper, e.g., to 25 cm, the soil is still warming from heat stored above that depth, at least initially. Working with Schmidt's method can provide interesting surprises at times. Information like this might help to explain the depth at which some animals choose to nest. Finally, although Schmidt's method is usually taught as a graphical procedure, it is easy to set up the same calculations in a spreadsheet, and to let it do the work.

11.4   Atmospheric Diffusion in Three Dimensions

The heat and mass balances we've studied so far have involved variations along a single spatial dimension, x. We now add variation with

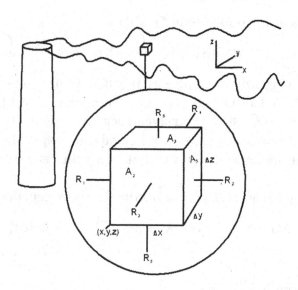

Figure 11.5: Spatial relationships involved in diffusion and advection of sulfur from a smokestack, with a cubic control volume (CV) as shown in the inset. West is to the left, and east to the right. Definitions: $V = \Delta x \Delta y \Delta z$, $A_1 = \Delta y \Delta z$, $A_2 = \Delta x \Delta z$, and $A_3 = \Delta x \Delta y$.

y and z, and also make two other changes. First, we allow for a bulk flow of air (wind) in addition to diffusive transport. Also, we allow for turbulent (eddy) diffusion in place of the simple molecular diffusion that results from Brownian movement of molecules.

Consider the small control volume shown in the plume from an effluent stack as shown in Fig. 11.5. We will derive the mass-balance equation for total sulfur as the plume from the stack moves along with a wind of speed U [m s^{-1}], which we take to flow in the positive x direction. Let x, y, and z, each with units of meters, be defined as shown in the diagram. The zero point is the top of the stack. Let S be the mass of sulfur [mg] present in the CV and $C = S/V$ be the sulfur concentration there [mg m^{-3}]. Note that C varies in both space and time, which is sometimes indicated by writing $C = f(x, y, z, t)$. At a particular point, its dependence on time is often indicated by the notation $C(t)$, while its dependence on location at a particular time is sometimes denoted by $C = f(x, y, z)$.

We define a given transport rate to be positive for material moving in a positive x, y, or z direction. A mass balance for sulfur in the CV over some short period of time Δt is thus

$$S(x, y, z, t + \Delta t) \approx S(x, y, z, t) + (\Sigma \text{ rates in } - \Sigma \text{ rates out})\Delta t.$$

We are working with an imaginary box (CV) whose front lower left corner is located at the point (x, y, z). From here on I will drop that cumbersome notation, but you should bear in mind that the notation $S(t)$ really means $S(x, y, z, t)$—the t is indicated explicitly to emphasize that we are interested in variation of S with t at a fixed point in space.

Referring to Fig. 11.5, we can rewrite the above equation as

$$S(t + \Delta t) \approx S(t) + (R_1 + R_3 + R_5 - R_2 - R_4 - R_6)\Delta t,$$

with units

$$\text{mg} = \text{mg} + \left[\frac{\text{mg}}{\text{s}}\right] \text{s}.$$

Because $C = S/V$, we now divide through by V and subtract $C(t)$ from both sides to obtain

$$C(t + \Delta t) - C(t) \approx \frac{(R_1 + R_3 + R_5 - R_2 - R_4 - R_6)}{V} \Delta t.$$

Dividing by Δt and taking the limit as $\Delta t \to 0$ yields

$$\frac{\partial C}{\partial t} = \frac{R_1 + R_3 + R_5 - R_2 - R_4 - R_6}{V}.$$

Next we need to express the six rates in more detail. First, R_1 and R_2 represent *bulk transport*, or *advection*, of sulfur (S carried along by the wind). This bulk transport of sulfur is very much like the mercury carried into and out of our famous lake by the streams that feed and drain it. In the latter situation the mercury transfer rate was of the form $\dot{m} = qC$ ([mg day^{-1}] = [L day^{-1}][mg L^{-1}]), where q was the stream discharge and C the mercury concentration. In the present case we have $R = AUC$ ([mg s^{-1}] = [m^2][m s^{-1}][mg m^{-3}]), so the air flow across the face of the box is A times the velocity U [m s^{-1}]. Thus

$$R_1 = A_1 U C(x) \text{ and } R_2 = A_1 U C(x + \Delta x).$$

The remaining Rs involve eddy diffusion. This is a process analogous to molecular diffusion, except that gas molecules are carried from regions of higher concentration to regions of lower concentration by eddies (gusts or packets of air) rather than by Brownian movement at the molecular level. Thus, eddy diffusivities (K) tend to be orders of magnitude higher than molecular diffusivities.

For these diffusion rates, we have that

$$R = -KA\frac{\partial C}{\partial \xi},$$

where ξ denotes any spatial dimension like x or y or z. K, the *eddy diffusivity*, has units of m^2 s^{-1}, so the whole term has rate units of

$$\left[\frac{m^2}{s}\right] m^2 \left[\frac{mg}{m^4}\right] = \frac{mg}{s}.$$

From this, we calculate that

$$R_3 = -K_y A_2 \frac{\partial C}{\partial y}\bigg)_y \quad \text{(S face)}$$

$$R_4 = -K_y A_2 \frac{\partial C}{\partial y}\bigg)_{y+\Delta y} \quad \text{(N face)}$$

$$R_5 = -K_z A_3 \frac{\partial C}{\partial z}\bigg)_z \quad \text{(bottom face)}$$

$$R_6 = -K_z A_3 \frac{\partial C}{\partial z}\bigg)_{z+\Delta z} \quad \text{(top face)}$$

where K_y is the horizontal diffusivity in the y direction (perpendicular to the wind) and K_z is the vertical diffusivity.

Now we substitute these detailed expressions for the Rs into the overall mass-balance equation to obtain:

$$\frac{\partial C}{\partial t} \approx \frac{1}{V}\left\{\left[A_1 UC(x)\right] + \left[-K_y A_2 \frac{\partial C}{\partial y}\right)_y\right] + \left[-K_z A_3 \frac{\partial C}{\partial z}\right)_z\right]$$

$$- \left[A_1 UC(x+\Delta x)\right] - \left[-K_y A_2 \frac{\partial C}{\partial y}\right)_{y+\Delta y}\right] - \left[-K_z A_3 \frac{\partial C}{\partial z}\right)_{z+\Delta z}\right]\right\}.$$

The next step is to collect together the x-ward, y-ward, and z-ward terms, factoring out constants where possible. We also substitute $V = \Delta x \Delta y \Delta z$, $A_1 = \Delta y \Delta z$, $A_2 = \Delta x \Delta z$, and $A_3 = \Delta x \Delta y$:

$$\frac{\partial C}{\partial t} = \frac{1}{\Delta x \Delta y \Delta z}\left\{ \Delta y \Delta z U\left[-C(x + \Delta x) + C(x)\right]\right.$$

$$+ \Delta x \Delta z K_y \left[\left(\frac{\partial C}{\partial y}\right)_{y+\Delta y} - \left(\frac{\partial C}{\partial y}\right)_{y}\right]$$

$$\left.+ \Delta x \Delta y K_z \left[\left(\frac{\partial C}{\partial z}\right)_{z+\Delta z} - \left(\frac{\partial C}{\partial z}\right)_{z}\right]\right\}.$$

Cancelling factors of Δx, Δy, and Δz yields

$$\frac{\partial C}{\partial t} \approx U\left[\frac{-C(x + \Delta x) + C(x)}{\Delta x}\right] +$$

$$K_y\left[\frac{\left(\frac{\partial C}{\partial y}\right)_{y+\Delta y} - \left(\frac{\partial C}{\partial y}\right)_{y}}{\Delta y}\right] + K_z\left[\frac{\left(\frac{\partial C}{\partial z}\right)_{z+\Delta z} - \left(\frac{\partial C}{\partial z}\right)_{z}}{\Delta z}\right].$$

Finally (whew!) we take limits as Δx, Δy, and Δz each go to zero, and obtain

$$\frac{\partial C}{\partial t} = -U\frac{\partial C}{\partial x} + K_y\frac{\partial^2 C}{\partial y^2} + K_z\frac{\partial^2 C}{\partial z^2}. \tag{11.7}$$

This is known as an *advection-dispersion equation*. The first-derivative term describes the advection (material carried by a bulk flow), and the second-derivative terms describe dispersion (or diffusion). We are assuming that x-ward diffusion is negligible relative to the x-ward advection. Solving Eqn. 11.7 would require us to choose some system boundaries, to set boundary conditions on that system at all times, and to set initial conditions (for $t = 0$) at every point within the volume—the solution process is beyond the scope of this book. Note that the "Gaussian plume" model, which you may meet elsewhere, results from solving this equation under certain simplifying conditions.

PDEs in Cylindrical Systems

So far we have worked with PDEs in rectangular coordinate systems, but some problems arise in cylindrical or spherical objects. In such cases it is usually much easier to work with differential equations in

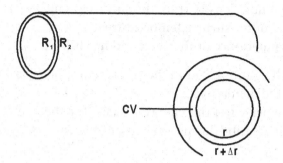

Figure 11.6: Diagram defining variables for modelling heat transfer along a cylindrical wire of radius R_1, with an insulation layer having an outer radius of R_2. The end view shows a control volume (CV); that view could also represent a section through the center of a sphere.

terms of cylindrical or spherical coordinate systems[4], which we considered earlier in Chapter 8. We will take up one simplified example of each.

We begin with a cylindrical case. The most general cylindrical coordinates allow for the quantity of interest (temperature, concentration, etc.) to vary with distance z along the length of the cylinder, with radius r out from the centerline of the cylinder, and with angle of rotation ϕ about the centerline. We will work with a simple special case in which temperature varies only with r (and time t), but not with z or with ϕ. This lack of angular dependence gives the problem *radial symmetry*.

Consider a long, round, copper electrical wire surrounded with a layer of plastic insulation, as shown in Fig. 11.6. We allow for heat production $G(t)$ [J cm^{-3} s^{-1}] caused by Joule heating; i.e., by an electrical current flowing through the wire. (Similar mathematics might apply to the elongated body of a weasel or a dachshund, with fur replacing the plastic insulation.) We will assume that heat is lost from the outer surface of the insulation according to

$$q = h[T(R_2) - T_e],$$

where

$$h = \text{convective heat transfer coefficient [J cm}^{-2}\text{ s}^{-1}\text{ C}^{-1}]$$

[4]An excellent description of various coordinate systems appears in Arfken (1970)

q = heat flux density from the surface [J cm^{-2} s^{-1}]

$T(R_2)$ = outer surface temperature [C]

T_e = temperature of the surroundings [C].

Then, we will be interested in the temperature profiles $T(r, t)$ in the wire and in the insulation.

In the copper wire, the energy balance for the control volume, a small cylindrical tube down the length of the wire, is, over some time period Δt:

> Heat content at $t + \Delta t$=
>
> Heat content at t
>
> + Heat conducted in at r
>
> + Heat generated within the CV
>
> − Heat conducted out at $r + \Delta r$.

Thus for a wire of length L [cm],

$$H(t + \Delta t) \approx H(t) + \left[-kA_r \frac{\partial T}{\partial r} \right)_r \right] \Delta t + GV\Delta t$$

$$- \left[-kA_{r+\Delta r} \frac{\partial T}{\partial r} \right)_{r+\Delta r} \right] \Delta t,$$

with units

$$J = J + \left[\frac{J}{cm \, s \, C} \right] \left[cm^2 \right] \left[\frac{C}{cm} \right] [s] + \left[\frac{J}{cm^3 \, s} \right] \left[cm^3 \right] [s]$$

$$- \left[\frac{J}{cm \, s \, C} \right] \left[cm^2 \right] \left[\frac{C}{cm} \right] [s].$$

Here V is the volume of the CV, and A_r and $A_{r+\Delta r}$ are its inner and outer surface areas, respectively. Expressing these areas and the volume in terms of L, r, and Δr yields $A_r = 2\pi rL$, $A_{r+\Delta r} = 2\pi(r+\Delta r)L$, and $V \approx 2\pi rL\Delta r$.

Strictly speaking, $V = \pi(r + \Delta r)^2 L - \pi r^2 L$, or $V = \pi L[r^2 + 2r\Delta r + (\Delta r)^2 - r^2]$. However, if we make Δr small enough (as we will in the limit), then the Δr^2 term will be negligibly small compared with $2r\Delta r$. Using these values, we can rewrite the energy balance as:

$$H(t + \Delta t) - H(t) \approx -k(2\pi rL) \frac{\partial T}{\partial r} \right)_r \Delta t + G(2\pi rL\Delta r)\Delta t$$

$$- \left[-k[2\pi(r + \Delta r)L] \frac{\partial T}{\partial r} \right)_{r+\Delta r} \right] \Delta t.$$

If we divide through by $c_p \rho V$ and by Δt, we have:

$$\frac{T(t+\Delta t) - T(t)}{\Delta t} \approx \frac{1}{c_p \rho} \left[-\frac{kr}{r\Delta r} \frac{\partial T}{\partial r} \right)_r + G \tag{11.8}$$

$$+ \frac{k(r+\Delta r)}{r\Delta r} \frac{\partial T}{\partial r} \right)_{r+\Delta r} \right],$$

with units of

$$\left[\frac{\deg}{s} \right] = \left[\frac{\text{g deg cm}^3}{\text{J g}} \right] \times \left\{ \left[\frac{\text{J cm}}{\text{cm deg s cm}^2} \right] \left[\frac{\deg}{\text{cm}} \right] + \left[\frac{\text{J}}{\text{cm}^3 \text{ s}} \right] \right.$$

$$\left. + \left[\frac{\text{J cm}}{\text{cm deg s cm}^2} \right] \left[\frac{\deg}{\text{cm}} \right] \right\}.$$

In the limit as $\Delta t \to 0$, Eqn. (11.8) becomes

$$\frac{\partial T}{\partial t} \approx \frac{1}{c_p \rho} \left[k \frac{(r+\Delta r) \frac{\partial T}{\partial r} \right)_{r+\Delta r} - r \frac{\partial T}{\partial r} \right)_r}{r\Delta r} + G \right].$$

We now would like to write the conduction term in the form of an appropriate derivative, but the first term on the RHS of this equation is not of the simple form

$$\lim_{\Delta r \to 0} \frac{\frac{\partial T}{\partial r} \right)_{r+\Delta r} - \frac{\partial T}{\partial r} \right)_r}{\Delta r} = \frac{\partial^2 T}{\partial r^2}.$$

However, to obtain a derivative, all we need in general is an expression of the form

$$\frac{f(r+\Delta r) - f(r)}{\Delta r},$$

which becomes $\partial f / \partial r$ in the limit as $\Delta r \to 0$. So all is not lost—we *define*

$$S(r) \stackrel{\text{def}}{=} r \frac{\partial T}{\partial r}.$$

With this definition, the conduction term becomes

$$k \left[\frac{(r+\Delta r) \frac{\partial T}{\partial r} \Big|_{r+\Delta r} - r \frac{\partial T}{\partial r} \Big|_r}{r\Delta r} \right] = \frac{k}{r} \frac{S(r+\Delta r) - S(r)}{\Delta r}.$$

In the limit, this reduces to

$$\frac{k}{r}\frac{\partial S}{\partial r} = \frac{k}{r}\frac{\partial}{\partial r}\left(r\frac{\partial T}{\partial r}\right)$$

$$= \frac{k}{r}r\frac{\partial^2 T}{\partial r^2} + \frac{k}{r}\frac{\partial r}{\partial r}\frac{\partial T}{\partial r} = k\frac{\partial^2 T}{\partial r^2} + \frac{k}{r}\frac{\partial T}{\partial r}.$$

Substituting this term back into our DE then yields

$$\frac{\partial T}{\partial t} = \frac{1}{c_p \rho}\left[k\left(\frac{\partial^2 T}{\partial r^2} + \frac{1}{r}\frac{\partial T}{\partial r}\right) + G\right].$$

This equation describes unsteady (time-dependent) conductive heat transfer in a cylinder, when temperature varies only in the radial direction. (The $1/r$ factor on the RHS is a general indicator of cylindrical geometry in problems like this.) If there were variations of G with z and with ϕ, or if the boundary conditions depended on these coordinates, then our equation would become (Kreith 1973)

$$\frac{1}{a}\frac{\partial T}{\partial t} = \frac{\partial^2 T}{\partial r^2} + \frac{1}{r}\frac{\partial T}{\partial r} + \frac{1}{r^2}\frac{\partial^2 T}{\partial \phi^2} + \frac{\partial^2 T}{\partial z^2} + \frac{G}{k}$$

where $a = k/(c_p\rho)$. We will not derive this more general form, or work with any problems that require it, but I present it here for your general information.

Spherical Coordinates

The most general *spherical* coordinate system also involves three coordinate variables, representing r, the distance out from the center of the sphere; ϕ, which is like longitude on a globe; and θ, which is similar to latitude. These were diagrammed earlier in Fig. 8.4, p. 179. As with the cylindrical example above, we will work only with a simple case where we allow for no angular variation of temperature, either ϕ-ward or θ-ward. This leaves T to vary only with r and t.

Look again at the circular inset in Fig. 11.6, p. 267. Previously it represented a cross section of a cylinder, but now we interpret it as cutting through the center of a sphere. This might represent the body of a small mammal rolled up into a ball, in which case heat production $G(t)$ [J cm^{-3} s^{-1}] would result from the animal's metabolism. For an application using a model like this for understanding the physiology of warm-blooded mammals, see Porter, Parkhurst, and McClure (1986).

The energy balance for the CV would again be of the form

$$\frac{\partial T}{\partial t} \approx \frac{1}{c_p \rho V} \left\{ k \left[A_{r+\Delta r} \frac{\partial T}{\partial r} \right)_{r+\Delta r} - A_r \frac{\partial T}{\partial r} \right)_r \right] + GV \right\}.$$

The expressions for areas and volumes now represent the spherical geometry:

$$A_r = 4\pi r^2; \quad A_{r+\Delta r} = 4\pi (r + \Delta r)^2; \quad V \approx 4\pi r^2 \Delta r.$$

Thus

$$\frac{\partial T}{\partial t} \approx \frac{k}{c_p \rho} \frac{4\pi (r + \Delta r)^2 \frac{\partial T}{\partial r}\Big)_{r+dr} - 4\pi r^2 \frac{\partial T}{\partial r}\Big)_r}{4\pi r^2 \Delta r} + \frac{G}{c_p \rho},$$

or

$$\frac{\partial T}{\partial t} \approx \frac{1}{c_p \rho} \left[\frac{k}{r^2} \frac{S(r + \Delta r) - S(r)}{\Delta r} + G \right],$$

where now

$$S(r) \stackrel{\text{def}}{=} r^2 \frac{\partial T}{\partial r}.$$

In the limit we have

$$\frac{\partial T}{\partial t} = \frac{1}{c_p \rho} \left[\frac{k}{r^2} \frac{\partial S}{\partial r} + G \right].$$

Now we carry out the differentiation of the new S:

$$\frac{\partial S}{\partial r} = \frac{\partial}{\partial r} r^2 \frac{\partial T}{\partial r} = r^2 \frac{\partial^2 T}{\partial r^2} + 2r \frac{\partial T}{\partial r}.$$

Finally, we put all this together and end up with

$$\frac{\partial T}{\partial t} = \frac{1}{c_p \rho} \left[k \frac{\partial^2 T}{\partial r^2} + \frac{2k}{r} \frac{\partial T}{\partial r} + G \right] = \frac{1}{c_p \rho} \left[\frac{k}{r^2} \frac{\partial}{\partial r} \left(r^2 \frac{\partial T}{\partial r} \right) + G \right]. \quad (11.9)$$

This equation describes conduction of heat in a sphere, with radial symmetry in the heat source and in the surface boundary conditions. It would apply to a rolled-up animal only if its surface heat loss were uniform around its whole body. If that heat loss varied from one point to another on its surface, then all the cylindrical coordinates would have to be brought in. The result would be:

$$\frac{1}{\alpha}\frac{\partial T}{\partial t} = \frac{1}{r^2}\frac{\partial}{\partial r}\left(r^2\frac{\partial T}{\partial r}\right) + \frac{1}{r^2\sin\phi}\frac{\partial}{\partial\phi}\left(\sin\phi\frac{\partial T}{\partial\phi}\right)$$

$$+ \frac{1}{r^2\sin^2\phi}\frac{\partial^2 T}{\partial\theta^2} + \frac{G}{k}.$$

As you can imagine, this full equation would be more difficult than Eqn. 11.9 to solve.

11.5 Exercises

1. Differentiate Eqn 11.1 once with respect to N to obtain Eqn 11.2, p. 252.

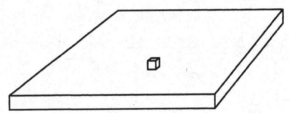

Figure 11.7: Geometry of a city located in a rectangular basin. The small cube in the middle is a control volume, with dimensions Δx, Δy, and Δz.

2. A city is located in a basin (Fig. 11.7) between a mountain range and some associated foothills[5]. The city is conveniently square (3 km on a side). One windy day, the SO_2 concentration in the air throughout the basin builds up to 10 mg liter^{-1}. Then suddenly the wind stops, and an inversion sets in at 100 m above the surface. The basin is surrounded by cliffs 100 m high (strangely enough).

Assume SO_2 production at any point on the surface is known in terms of a function $P(x, y, t)$, in kg km^{-2} hr^{-1}. (P would be high at a coal-fired power plant, and might be negative at a location

---

[5]This problem is somewhat oversimplified to make it easier to deal with. Even so, you may find that it is not *too* easy.

where plants absorbed SO_2.) The physics of diffusion across a plane, which is similar to the physics of heat conduction, can be described as:

$$R = -KA\frac{\partial C}{\partial x},$$

where R is diffusion rate across a vertical plane of area A [km^2], in kg hr^{-1}, K = diffusivity of SO_2 in air, [km^2 hr^{-1}], C = SO_2 concentration, [kg km^{-3}], and x = distance in the direction perpendicular to the plane. [km]

In any small control volume, the SO_2 concentration C at time $t + \Delta t$ will approximately equal $C(t)$ plus

$$\frac{\text{Net mass of } SO_2 \text{ entering control volume during } \Delta t}{\text{Volume of control volume}}$$

because concentration is mass per unit volume.

Your tasks are to:

a) Identify the 6 terms that define the net rate of SO_2 accumulation in the volume. For each of these terms, you will need to multiply a rate per unit area (kg km^{-2} hr^{-1}) times the area across which that transfer is occurring to obtain a rate in kg hr^1.

b) Add the terms that add SO_2 to the volume, and subtract the terms that take SO_2 out. Be careful though, since some of the terms have their own minus signs from Fick's law.

c) Multiply this net rate (from B) times Δt to obtain the approximate net increase in kg of SO_2 during the time increment.

d) Divide the result from C by the volume of the control volume (in km^3), to get the net increase in SO_2 *concentration*.

e) Rearrange terms to get time dependence on one side and space dependence on the other. Also, group x-dependent, y-dependent, and z-dependent terms separately.

f) Take limits as Δx, Δy, Δz, and Δt each go to zero, and you have your PDE.

The initial condition is that $C(x, y, 0) = 10$ mg L^{-1} = 10^7 kg km^{-3} for all x, y, and z. The *boundary conditions* are that $\partial C / \partial x = 0$

on the east and west boundaries, and $\partial C/\partial y = 0$ on the north and south boundaries. These conditions are consistent with zero diffusion through the cliffs that bound the city. Also, $\partial C/\partial z = 0$ at the inversion ($z = 0.1$ km), but

$$-K \left. \frac{\partial C}{\partial z} \right]_{x,y} = P(x,y,t) \text{ at } z = 0.$$

3. Consider a solar heat collector made of a copper sheet that is H mm thick by W cm wide by L cm long. The thickness H is small enough that temperature differences perpendicular to the plane of the sheet may be neglected. That is, no z component of heat transfer needs to be considered unless you wish to do so. The copper has conductivity k (J cm^{-1} deg^{-1} sec^{-1}), density ρ (g cm^{-3}), and specific heat c_p (J g^{-1} deg^{-1}).

Heat is absorbed by the upper surface of the sheet at a rate that (for practical purposes) depends only on the current solar radiation striking the plate. This rate, $S(t)$ [in J cm^{-2} sec^{-1}], varies with time but is uniform over the surface of the plate (i.e., it is independent of the x and y coordinates).

Heat is transferred out of the lower surface of the sheet to water that flows through tubing soldered to that surface, at a rate that varies with x, y, and the local temperature $T(x,y,t)$. Specifically, this rate $[Q(x,y,t)]$ is given by

$$Q = b(x,y) \times [T(x,y,t) - \theta(x,y,t)]$$

where b is an exchange coefficient (J cm^{-2} sec^{-1} deg^{-1}), and θ is the water temperature in the nearest tubing (deg).

In summary, heat can be transferred into and out of any small area on the plane by x-ward conduction, y-ward conduction, adsorption of solar energy on the top, and transfer to the water in the tubing on the bottom.

Your assignment is to derive the partial differential equation that summarizes the net heat balance at every point on the plane of the copper sheet. For this derivation, start with a control volume of thickness H that runs from x to $x + \Delta x$, and from y to $y + \Delta y$.

4. The core of a nuclear power reactor at Goldurnsk is a sphere of radius R in which, after the graphite rods are removed, the rate of heat production is uniform at Q [J cm^{-3} min^{-1}]. The outside of the core is kept at a uniform temperature T_R by heat-exchanging fluid (usually liquid mercury); this ensures that the temperature within the core will vary with radius only, and not with "latitude" or "longitude".

   Derive the differential equation whose solution would allow calculation of the core temperature at any distance r from the center and at any time t after the graphite rods are withdrawn. Ignore the hollow spaces left by the rods, and assume the remaining core material is uniform in thermal conductivity k [J cm^{-1} min^{-1} deg^{-1}], density ρ [g cm^{-3}], and specific heat c [J g^{-1} deg^{-1}].

5. Migration of plant and animal populations is in some ways analogous to molecular diffusion, at least if animals tend to move from areas of higher density to areas of lower density in proportion to population density gradients. Some mathematical biologists (e.g., Holmes, Lewis, Banks, and Veit 1994) have written models that exploit this analogy, as we do here.

   Suppose two rabbits (that's all it takes if you pick the right two) are plunked down at a point in the middle of the Central Australian desert. There they begin to reproduce, and to migrate outward in a circular pattern. Assume that:

   • Birth and death processes on any small area of land follow a logistic model of the form

   $$\frac{\text{Births} - \text{Deaths}}{\text{time}} = rN\frac{K - N}{K}$$

   where births and deaths are in units of rabbits, time is in years, and r = relative net reproductive rate [rabbits/rabbit/year], N = rabbit *density* [rabbit/m^2], and K = carrying capacity [rabbits/m^2].

   • The rabbits migrate from more crowded to less crowded areas, in a process somewhat like molecular or eddy diffusion. In particular, their rate of movement across any L meters of line is given by

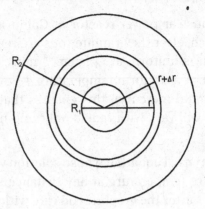

Figure 11.8: A circular coordinate system.

$$Q = -mL\frac{\partial N}{\partial r},$$

where Q = migration rate [rabbit/year], m = "migrativity" [m^2/year], L = length of line across which migration is taking place [m], and $\partial N/\partial r$ = gradient of rabbit density in the r direction, [(rabbits/m^2)/m]=[rabbits/m^3].

Your job is to write the partial differential equation for N as a function of r and t, assuming a circular coordinate system centered on the point of release. Use a "control area" that is a narrow ring of area as shown in Fig. 11.8.

Two hints:

• The diffusive movement of rabbits into the control area from the inside (at r) takes place across a line of length $2\pi r$, while the movement outward from the control area occurs across a line of. length $2\pi(r + \Delta r)$.

• Do not attempt to write separate DEs for the birth and death versus the migration processes. Migration from an area responds to *total* density in that area, regardless of whether the animals were born in the area or migrated into it. Similarly, rabbits in a given area can reproduce and die regardless of whether they were born there or migrated into that area. Thus, your one PDE should combine both processes.

6. An entomologist who studies insects that bore into giant sequoia

(*Sequoiadendron giganteum*) trees wants to know how the temperature varies with time and depth in the trunk, as daily warming and cooling occurs at the tree's surface. If we treated the trunk as a circular cylinder, the appropriate equation for the energy balance would be:

$$\frac{\partial^2 T}{\partial r^2} + \frac{1}{r}\frac{\partial T}{\partial r} = \frac{1}{a}\frac{\partial T}{\partial t}.$$

It is possible to modify Schmidt's method to allow its use in solving this equation, but we won't add that complication here. Instead, if a cylinder is large, then temperature variations near its surface will not be very different from those in a solid with a flat surface. In such a case, we can approximate the energy balance above with

$$\frac{\partial^2 T}{\partial x^2} \approx \frac{1}{a}\frac{\partial T}{\partial t},$$

where x is the depth from the surface into the trunk. You have seen how to apply Schmidt's graphical method to solving that equation.

Now, suppose that the surface temperature $T_s(t)$ on the north side of a "big tree" trunk during some midsummer period varies approximately as

$$T_s(t) = \bar{T} + b\,\sin[c(t - d)],$$

where $\bar{T} = 21$ C is the mean temperature (over several days), $b = 3$ C is the amplitude of the sinusoidal variation, $c = (2\pi/24)$ hr^{-1} adjusts the period to 24 hr, and $d = 9$ hr is the "phase shift" that causes the maximum to fall at 3 p.m. This equation assumes that midnight of the starting day is equated to $t = 0$.

Use Schmidt's method to study the penetration of the daily heat wave into one of these large trees. As an initial condition, assume that the temperature is uniform at $\bar{T}$, at a starting time of 9 a.m. one morning. Then, let the surface temperature vary sinusoidally as given.

Kreith (1965) gives a value $a = 0.0029$ ft^2 hr^{-1} for pine when the heat transfer is perpendicular to the grain as it is here. Convert that value to cm^2 hr^{-1}, and estimate graphically the changes in temperature in the tree trunk for eight time steps, using ten Δx steps of 2 cm each. The surface temperature climbs for eight time steps, and then begins to decline again. The process continues to work in that case, but the lines start to overlap confusingly unless you switch to a new sheet of paper. Assume (probably wrongly) that the thermal properties of the bark, as represented by a, are the same as those in the wood. Schmidt's method can be adjusted for varying thermal properties, but we won't deal with that here. Also assume that the temperature at 20-cm depth remains at $\bar{T}$ for the duration of your calculations.

Once you have completed those manipulations, plot the temperature at 2 cm depth as it varies with time. This might be of interest to the entomologist, if that were a typical depth to which the insects bored.

Coda: I once used thermocouples to measure temperatures every hour for 24 hr, at 2 cm intervals in a 20-cm diameter Ponderosa pine tree in Colorado. Interestingly, the temperature at the center of the tree (at a 10 cm depth) reached a maximum at a time very close to 12 hours after the surface temperature reached its maximum. As a related point, the mass of wood present in a forest acts as a sort of "thermal sponge," and reduces temperature variations relative to what they would be in the same climate if the trees weren't there. Wood is not as effective as water in this regard because its heat capacity is lower, but it can still moderate temperature variations substantially.

7. A wastewater treatment plant has a long, rectangular primary settling basin that is L m long, W m wide, and H m deep in the x, y, and z dimensions, respectively. Water moves in the x direction, from 0 to L; the flow rate is q m^3 s^{-1}. The water entering the tank at $x = 0$ contains a concentration C_u of solids; this concentration varies with time, so we can think of $C_i = C_i(t)$ [g m^{-3}]. (This is the boundary condition. The subscript i indicates input to the upstream end of the basin.)

As the water moves down the basin, the solids settle out, which is the main purpose of the structure. Specifically, within any small volume, a fraction p of the solids drop to the bottom each second (on an instantaneous basis). Your task is to derive the rate equation whose solution would tell us the concentration of solids in the water at any point x and at any time t. Describe in general terms what the form of the initial condition would be for a problem like this. You do not have to solve the equation. You should make the simplifying assumptions that both the water flow and the concentration of solids are uniform across any vertical cross section of the basin (i.e., that neither varies with y or with z.)

Note that this situation involves only advection of the solids, and not diffusion. Provide unit checks for the first equation you write down (whatever form that takes) and for your final answer.

8. One of the many uncertainties related to global warming is the possibility of positive and negative feedbacks that may exacerbate or ameliorate the general effect. For example, as warming occurs in northern latitudes, organic matter that has accumulated from slow decomposition caused by low temperatures may rot faster, releasing CO_2 and methane in the process. These gases could then "magnify" the greenhouse effect caused originally by burning fossil fuels.

Geochemists find, not surprisingly, that CO_2 and methane emission rates increase with temperature in the peat. Suppose you are working with a research team who wish to model heat flow within a peat bog. Your task here is to derive the partial differential equation describing heat conduction in a region of a bog where the following simplifying conditions hold:

• Temperature may vary with time, depth z, and both horizontal coordinates (x and y). Express distances in cm.

• The peat is fairly homogeneous, with uniform thermal conductivity k [J cm^{-1} s^{-1} deg^{-1}], density ρ [g cm^{-3}], and heat capacity c_p [J g^{-1} deg^{-1}].

• No water is flowing, so heat moves only by conduction (opposite to temperature gradients in the way we have studied).

• Heat is produced by the decomposition process at a temperature-dependent rate $f(T)$ [J cm^{-3} s^{-1}].

Because no shape has been specified for the region of interest, you need not be concerned with boundary or initial conditions, and solving the equation is out of the question. However, show your derivation in detail.

9. As you have seen, Schmidt's method allows easy but approximate solution of the equation for transient (time-dependent) conduction of heat in one physical dimension; i.e.,

$$\frac{\partial T}{\partial t} = a \frac{\partial^2 T}{\partial t^2},$$

where $a = k/(c\rho)$ is the thermal diffusivity of the medium. A gas pipeline is currently under consideration to be constructed next to the Alyeska oil pipeline that crosses part of Alaska (this is not just hypothetical—it may have already been constructed). Suppose you work for a consulting firm, and that you are assigned the task of calculating pre-construction temperature variations in the soil profile near where the proposed pipeline is to run. At your starting point, the temperature profile (temperature T versus depth z) has the form shown here:

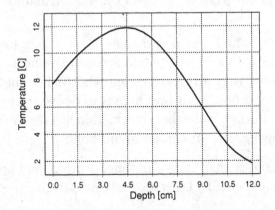

You propose to your supervisor to use Schmidt's method, and you show her an example of how it works. She finds it interesting, but says that you need to use calculations rather than a graphical method to allow use of more, smaller Δz steps, smoother plotting, and numerical results to be used in calculations of other effects. You respond "Aha!" Recalling that the points you plot in the graphical method are simply averages of two earlier values, you realize how easy it would be to do calculations that are completely equivalent to the Schmidt manipulations.

Consider a soil for which $a = 36$ cm^2 hr^{-1} (van Wijk 1963). Starting with a temperature profile given by the values in the $t = 0$ row of the table below (and shown in the figure above), and given the boundary conditions provided in the $z = 0$ and $z = 12$ cm columns, fill in the table for the next two time steps ($t = \Delta t$ and $t = 2\Delta t$). Provide the temperature estimates in a table like the one below. You may assume that no water is freezing or thawing during the time period you are considering.

Finally, determine and state (with units) what the numerical value of $t = \Delta t$ is, so one can associate each row of the table with a specific time.

	Depth z [cm]								
	0.0	1.5	3.0	4.5	6.0	7.5	9.0	10.5	12.0
t	Temperature [deg C]								
0	7.7	9.8	11.3	11.9	11.1	8.9	6.0	3.2	1.8
Δt	7.2								1.8
$2\Delta t$	6.7								1.8
...					...				

10. A factory uses an organic solvent, some of which evaporates from its manufacturing process and escapes into the outside air through a vent pipe in the roof. From there the vapors are distributed and destroyed by the following processes:

   a) They are carried along by a wind that always blows straight x-ward, from west to east, but with variable speed $v(t)$ [m s^{-1}].

   b) They diffuse vertically, by turbulent (eddy) diffusion, with diffusivity K_z [m^2 s^{-1}].

   c) They diffuse horizontally, northward and southward, with diffusivity K_y [m^2 s^{-1}].

   d) They are broken down photochemically into more harmless substances. In this process, a fraction f of the vapor in any small volume degrades each second. This f is not a constant; rather, it is proportional to solar intensity, which varies with time of day.

Using a small control volume, derive the partial differential equation which, if solved, would determine the concentration of solvent vapors at any point x, y, z relative to the outlet of the vent pipe as the origin.

11. Population biology has a rich tradition of mathematical modelling. As noted earlier, some population biologists add migration to their models by borrowing the diffusion concept from physics. We adopt that analogy here.

Consider a large area of relatively uniform forest. A new insect pest is introduced near the center of this area by the careless importation of some infested logs. The pest population grows and expands as follows. On any given hectare, its "natural" growth rate (birth rate minus death rate) is logistic, with that rate G [kg biomass ha^{-1} da^{-1}] being

$$G = \rho B \left(\frac{K - B}{K} \right).$$

Here ρ is the specific growth rate [da^{-1}], B is the biomass density [kg ha^{-1}], and K [kg ha^{-1}] is the short-term carrying capacity of the forest for the insect density.

In addition to this growth, the insects migrate radially (in ever-increasing circles) analogously to diffusion, along biomass-density gradients. Thus, across any line or line segment (including curved ones), the pests migrate at a rate M [kg per meter of line length per day] given by

$$M = -D \frac{\partial B}{\partial \omega},$$

where D is the insect "diffusivity" [m^2 da^{-1}] and ω is any coordinate, such as x or r, along which the pests are moving.

Your task is to derive the equation that, if solved, would allow prediction B's variation with time and with distance from the point of the original introduction. You need not attempt to solve the equation—just derive it. Fig. 11.8, p. 276, would be useful here; with the outer ring representing the outer boundary of the forest region, the point of contamination being the center of the circles, and the narrow ring being a control area. The area of the latter is approximately $2\pi r \Delta r$.

Be sure to check the units of your final result, noting that one hectare is 10^4 m^2. (For safety, you may wish to check units at several stages in your derivation.)

12. An insect pest is carelessly introduced more or less uniformly along a road at one edge of a rectangular strip of forest. (The short dimension of the rectangle is along the road, and the long dimension is perpendicular to the road.)

    The insects move into and then through the forest in a diffusive fashion, with a dispersal coefficient (a "migrativity") of M m^2 da^{-1}. They have a reproduction rate of b percent per day, and a fraction f are eaten by predators, or die for other reasons, each day.

    Derive the PDE that would yield as its solution the density D (in insects m^{-2}) at any time t and at any distance x into the forest. The x coordinate is measured perpendicular to the road. You need not attempt to solve this equation. Be sure to include at least one unit check.

11.6 Questions and Answers

1. What Greek letter is the partial derivative symbol?

   • It's not a Greek letter at all. I think it's Elvin, from the Fourth Realm <grin>. Actually, it may be somebody's idea of a curly 'd'. Just read it as 'partial' or, when you're not worried about confusion with ordinary derivatives, just read it as 'd'.

2. In the $G(N, P)$ example with algal growth, can you also get the partial of N with respect to P (and vice versa), and would you ever want to do that?

   • First, yes, you *can*. One way (for that particular function) is to solve for N as a function of G and P, and then take $\partial N / \partial P$. If it were a messier function, and you couldn't isolate the N on one side, you could still use a process called implicit differentiation to get $\partial N / \partial P$. See most any calculus text for details on that process.

   Second, would that be *meaningful*? Yes, in many situations, including this one. Here it would tell you about how much you would have to change N for a unit change in P, to keep G constant. It seems from looking at the plot that this derivative would therefore be negative.

3. You derived the finite difference scheme for a second *partial* derivative. Does that equation also apply for ordinary second derivatives? I'm referring to the $[y(x+h) + y(x-h) - 2y(x)]/h^2$ expression.

   • Yes. That same formula is sometimes used to get approximate (numerical) solutions to second-order *ordinary* DEs, too.

4. Please explain again why partial differential equations don't always involve second derivatives.

   • So far in Chapter 11 we've dealt with fluxes that were proportional to temperature or concentration *gradients*. The rate of diffusion across a plane is proportional to the first derivative of concentration. The rate of heat conduction across a plane is proportional to the first derivative of temperature. Since the first derivative can change from x to $x + \Delta x$, that leads to second derivatives when we shrink to a point.

   Not all fluxes are proportional to gradients, however. When a material, or for that matter, heat, is carried across a plane by a flow of some fluid (like water or air), the flux depends on the concentration (or temperature) in that fluid, not on the gradient. Now it's the concentration (temperature) itself that changes from x to $x + \Delta x$, so when we take a difference and shrink to a point, we get a *first* derivative.

Appendix A

Pre-Calculus Math Review

This section is provided for those wishing to review basic pre-calculus math.

Rules of Precedence

To be most valuable as a language of science, math needs to be as unambiguous as possible. For example, the following equation appeared in a statistics paper by Chen (1995):

$$\hat{\beta}_1 = \Sigma(X_i - \bar{X})^3/n/S^3. \tag{A.1}$$

It is not obvious whether that means

$$\hat{\beta}_1 = \frac{\Sigma(X_i - \bar{X})^3}{(n/S^3)} \quad \text{or} \quad \hat{\beta}_1 = \left[\frac{\Sigma(X_i - \bar{X})^3}{n}\right]/S^3 = \frac{\Sigma(X_i - \bar{X})^3}{(nS^3)},$$

which *are* different quantities. This equation arises in the "Chen test," which the EPA specifies for use at some hazardous waste sites (US EPA 1996), so it, like most equations, is important to "get right." The second form turns out to be correct, but it would have been helpful if the author had used parentheses and written the equation unambiguously to begin with as

$$\hat{\beta}_1 = \Sigma(X_i - \bar{X})^3/(nS^3). \tag{A.2}$$

It is useful to know the *rules of precedence* for performing mathematical operations, and then to use lots of parentheses to make the order even more clear. Those rules are that parentheses and their relatives[1]—(), [] or brackets, and {} or braces—act first, exponentiation (e.g., taking x to the power y) comes next, multiplication and division are next, and addition and subtraction come last. Also, operations at a given level (e.g., multiplication and division) are performed in order from left to right. Thus $a + b \times x = a + (b \times x)$, $a \times b + c^2 \times d = (ab) + [(c^2)d]$, and $a/b \times c = (a/b)c$. Note that by these standard rules, $a/b \times c$ is equal to $(ac)/b$, *not* to $a/(bc)$.

As another example, $1 + 2/3 + 4 = 1 + (2/3) + 4 = 5\frac{2}{3}$, not $\frac{3}{7}$. If you mean $(1+2)/(3+4)$, be sure to put in the parentheses. For another example, $3/5^2 = 3/(5^2)$, not $(3/5)^2$. As it happens, Chen's equation A.1 is correct if interpreted by these rules (check it out!) but even so, it would be safer to write it as Eqn. A.2 in the first place.

Incidentally, two common words that have specific meanings in mathematics are *term* and *factor*. In general, I will use "term" to mean an expression within a formula that is added to or subtracted from other terms, and "factor" to mean an expression that is multiplied by other factors within a term. For example, in the series

$$f(x) = f(a) + \frac{f'(a)}{1!}(x-a) + \frac{f''(a)}{2!}(x-a)^2 + \frac{f'''(a)}{3!}(x-a)^3, \quad \text{(A.3)}$$

which will appear in the next chapter, the expressions $f(x)$, $f(a)$, and $f''(a)(x-a)^2/2!$ are terms, while $f''(a)$, $(x-a)^2$, and $1/2!$ are factors within the fourth term. Although these two words are sometimes used more loosely, it is helpful to maintain these more precise meanings when discussing math.

Note that in Eqn. A.3, $f(x)$ denotes the value of the function f at some value of the independent variable x, and $f(a)$ denotes the value of the function at a particular point where $x = a$.

Some Algebraic Operations

Below is a list of relationships from algebra that we'll need from time to time in the book. I'd suggest practicing with them now, by way of review. To use these profitably, please consider the following points:

---

[1] Caution: brackets and braces are useful in mathematical text to help the eye find matching sets; however, most computer languages give special meanings to those symbols, and only ordinary parentheses are usually allowed to delimit sections of equations.

1. It's easy to confuse "levels of knowing"—it is one thing to watch someone else work through a relationship, and to say to yourself "I see that." It requires a different, much deeper level of understanding to be able to work through similar examples on your own[2]. This deeper understanding is what is needed for effective use of mathematics in environmental science, and it can be developed with practice. For this reason, I suggest that you work with the relationships below something like this:

2. Look at each relationship, and decide whether it is something you already know really well (as well as you know that "one plus one equal two," for example). If so, go on to the next one. If not, don't just look at it—instead, *work* with it. Put in some numbers. Demonstrate for yourself that the relationship is true, for at least a couple of cases.

Here's an example: Suppose the relationship given is $x^a x^b = x^{a+b}$, which shows one of the ways in which exponents combine. You could just try to memorize this, but you'll learn it much more effectively if you plug in some numbers, like $2^3 2^4 = 2^{3+4} = 2^7$. That looks reasonable, but to be sure, check it out by calculating 2^3, 2^4, and 2^7, and confirming whether it all checks out. As you can see, $8 \times 16 = 128$, so it does check. (Using your calculator to calculate 2^7 will give you practice with specialized calculations of that type, too, so this is an additional advantage you can gain from this sort of exercise.)

Working with relationships in this way, in contrast to just reading about them, not only gives you a deeper understanding, but also helps to imprint them in your memory. I suggest that you work with all the others in the list in similar ways.

1. $x^a x^b = x^{a+b}$

2. $e^{x+y} = e^x e^y$ (This is the same relationship, for a particular "x." As you probably know, e is the base of natural logarithms. Incidentally, e^x is often written as "$\exp(x)$." The "exp" is known as the *exponential function*.)

---

[2]Because of this, you are strongly advised not to look at the answers to problems until after you've done your best to solve them on your own

3. $x^{-p} = 1/x^p$ (the definition of a negative exponent)

4. $x^{a-b} = x^a/x^b$ (Check out how this follows from the relationships above.)

5. $(ax)^p = a^p x^p$ (and as a special case when $p = 0.5$, $\sqrt{xy} = \sqrt{x}\sqrt{y}$).

6. $x^{1/n} = \sqrt[n]{x}$ (It is *not* true that $x^{1/n} = 1/x^n$.)

7. $e^{\log x} = x$ and $\log e^x = x$. (These follow from the definition of a logarithm. What are the comparable expressions for base 10 logarithms?)

Properties of Logarithms

Because logarithms play an important role in many areas of applied math, it is worthwhile to review their properties here. By definition, the logarithm to the base b of a quantity x is simply the power (exponent) to which b must be taken to yield the quantity x. For example, 2 is the log (base 10) of 100, and 0.5 (or 1/2) is the log (base 10) of the square root of ten. In general, $b^{\log_b x} = x$; for example, $10^{\log_{10} x} = x$.

When the base is $b = 10$, we refer to *common logarithms*. When b is the number e (≈ 2.71828183), we refer to *natural* (or Napierian) *logarithms*. These are the two most frequently used bases, but base 2 can be useful when halving or doubling is of particular interest, and other bases are permitted if needed.

Notation can be a little confusing—in some scientific and engineering writing, $\log x$ refers to the common (base 10) log, and $\ln x$ refers to the natural (base e) log. However, "real mathematicians" often use "$\log x$" for natural logs, and "$\log_{10} x$" for common logs, because natural logs arise naturally in their applications, and common logs do not. Thus, when using various computer software (spreadsheets, statistical packages, languages like Fortran, and mathematical programs like MATLAB and MathCad) to perform calculations involving logs, be sure to check the appropriate names for the different logarithms. You also have to be careful when you read scientific papers and books. If you see "log" without a base subscript attached, you have to figure out from the context which base is intended. In this book, "log" will refer to the natural (base e) log; if base-10 logs are needed, they will be denoted by "$\log_{10}$."

The main properties of logarithms follow from their definition and from the properties of exponents. For example:

1. If $y = x^a$, then $\log y = a \log x$.

2. If $z = x \cdot y$, then $\log z = \log x + \log y$.

3. $\log 1/x = \log x^{-1} = -\log x$.

4. If $z = x/y$, then $\log z = \log x - \log y$.

5. $10^{\log_{10} x} = x$ and $\log_{10} 10^x = x$

6. $\log xy = \log x + \log y$ (These same relationships hold for logs to any base.)

7. $\log(x/y) = \log x - \log y$

8. $\log x^p = p \log x$

9. $\log(x + y) = \log(x + y)$!. Be careful—the log of a sum can't be simplified further. If you needed a numerical value, you would have to calculate the sum before taking its log.

10. $\log(x^a y^{-b})$ Based on the relationships above, try expressing this in terms of $\log x$ and $\log y$.

Now we get away from logarithms, and look at other relationships:

1. $(ax + by)^2 = (ax)^2 + 2axby + (by)^2$ Try showing this by multiplying out $(ax + by)(ax + by)$.

2. $(ax + by)^3 = (ax)^3 + 3(ax)^2 by + 3 ax(by)^2 + (by)^3$

3. $(ax + by)^4$ Try working this one out.

I. Without using a calculator, simplify to a single number:

1. $2^5 \times 2^3 = ?$;   2. $(3^2)^3 = ?$;   3. $(2/3)^2 = ?$;   4. $2^5/2^2 = ?$;

II. Some of these "equalities" are correct and some are not. Mark each as true or false:

1. $a^2 \times b^5 = (ab)^7$

1. $a^2 \times b^5 = (ab)^7$ 2. $\dfrac{1}{a} + \dfrac{1}{b} = \dfrac{a+b}{ab}$

3. $3a + 4b = 7ab$ 4. $\dfrac{a+b}{c+d} + \dfrac{1}{c+d} = \dfrac{a+b+1}{c+d}$

5. $x^2 + y^2 = (x+y)^2$ 6. $\dfrac{x+y}{x+z} = \dfrac{y}{z}$

7. $\dfrac{x}{y} + \dfrac{r}{s}$ 8. $\dfrac{x}{y} + \dfrac{r}{s} = \dfrac{sx+ry}{sy}$

9. $\dfrac{a}{b}x = \dfrac{ax}{bx}$ 10. $\left(\dfrac{ax+bx}{x+cx}\right)/x = \dfrac{a+b}{1+c}.$

III. Some important relationships:

1. What is the equation of the straight line passing through points (-1,1) and (2,16), where each of the points consists of (x,y) coordinates?

2. If $2x^2 + 3x = 5$, what is x?

3. Given $\log 2$ and $\log 5$, find $\log 10$ without using tables or calculator.

4. Given $\log_{10} 5$, find $\log_{10} 2$ without using tables or calculator.

Calculator Problem Set

These exercises are to help you ensure that you know how to use your calculator. It is to your advantage to be sure you can do all these operations—you may need to do similar calculations in exercises as you work through this book. Try not to do any subcalculations in your head (except to check results), and not to write down intermediate results if that can be avoided.

1. Simple operations: 5×3; $15/3$; $6 + 5 - 1$

2. $5 \times 3 \times 7$; $6 \times 4 + 2$; $3(6 + 5 + 4)$

3. $[(2+3) \times (4+5)]/[(6+7) \times (8+9)]$

4. $[(2)(3) + (4)(5)]/[(6)(7) + (8)(9)]$

5. exp 1; log 10; sin 3.14 (radians, not degrees)

6. exp(log 10); log[5(3) + 2]

7. Store $x = 3.3$ in a memory. Then calculate $(\{[(2x - 3)x + 4]x - 5\}x + 6)$. This yields the value of the polynomial $y = 2x^4 - 3x^3 + 4x^2 - 5x + 6$ at $x = 3.3$. (Be sure you see why this is so.) This technique for evaluating polynomials is computationally efficient, and it usually also produces less round-off error than evaluating the polynomial term by term. It is known as Horner's Rule.

8. Obtain the natural log of the previous answer without repunching that answer.

Answers:
2c: 45; 3: 0.20362; 4: 0.22807; 5a: e = 2.71828; 5b: 2.30259; 5c: 1.59265E-3; 6a: 10; 6b: 2.8332; 7: 162.4332; 8: 5.0903.

Functions

One basic and important concept in mathematics is that of the *function.* For example, $y = 3x + 4$ represents a linear (straight-line) relationship between the variables x and y. This function might often be written as $y = f(x) = 3x + 4$ as well, indicating that y, the *dependent variable,* depends on x, the *independent variable.* Note that although "3(4) = 12" uses parentheses to represent multiplication, in forms like $f(x)$ or $g(t)$, the parentheses indicate a function rather than multiplication. Thus, "$f(x)$" is read as "f of x," indicating that f is a function of the variable x. It does *not* mean f times x.

Among its many other uses, function notation allows compact statements of "what depends on what." Thus, "$T(x)$" might indicate that temperature T depends on (or varies with) distance x. Similarly, "$c(t)$" might indicate that some concentration c varied with time t. Be careful with units. Because $T(x)$ means "temperature *at* x" (not temperature *times* x), the units of $T(x)$ are just degrees, not degrees × cm (or some other length units).

If you have not used functional notation recently, please review the concept in an introductory calculus text. **It is critical that you recognize notation like f(t) as representing a function of t, and not as a quantity f multiplied by a quantity t.**

Appendix B

Solutions to Odd-Numbered Exercises

I recommend that you not look at answers until you have taken the associated exercises as far as you can. You will learn a thousand times more (give or take a little) by working a problem through yourself than by looking at a prepared solution and convincing yourself that you understand it. Remember that in your future work you will have to solve problems yourself, not just understand someone else's solutions (and the same is true in exams, if you are using this book in a course).

If you help others with an exercise, you can do them the most good not by working it for them, but rather by either (a) working some similar problem for them, or (b) determining where they are stuck, and asking leading questions that will get them on the right track. If you go to the effort to help others in these ways, you will likely learn a lot yourself in the process.

Chapter 1

1. 133.$\bar{3}$ mi
3. 1.4615 T
5. $\approx$42.795 m
7. $T_B \approx 13.379$ hr
9. 125 miles
11. L $\approx$ 76.430 + 0.00129412 T
13. $(1.4)^{2/3} = 1.2515$, so the increase is 25.14%.

15.

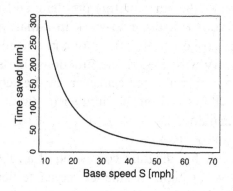

17. A.
$$T = \frac{-T_1 r_2 - T_2 r + T_2 r_1 + T_1 r}{-r_2 + r_1}$$

B. $T \approx 5.26°C$

19. Not enough information is given. If you obtained a numerical answer, then you must have made some assumption that can't be justified on the basis of what is given. The difficulty is that the costs might be different in the two five-year periods.

21. A. $a = 330$ and $b = 170$, both cal cm^{-2} day^{-1}; B. $c = 2\pi/364$ day^{-1}; $d = -81$ day

Chapter 2

1.
$$\frac{df}{dt} = -\frac{e^{-ct}}{gt^2 + k}\left[a \sin\left(a\left(t + b\right)\right) + c \cos\left(a\left(t + b\right)\right) + \right.$$

$$\left. 2\frac{gt \cos\left(a\left(t + b\right)\right)}{gt^2 + k} \right]$$

You might arrange the result in a different way, so yours could look different from this but still be correct. One way to check whether two forms are (probably) the same is to put numerical values into both, and to see whether they yield the same final result. I usually use a different prime number (2,3,5,7,11,13,...) for each constant or variable in an expression, to lower the chance of cancellation of errors. The numerical value requested is about 0.00752028531.

3. The numerical value of the analytic derivative is $-\sin(2)$, or about -0.9092974268. (Be sure for this problem, and throughout the book, to set your calculator to work in radian mode whenever you calculate a trigonometric function like sin, cos, tan, etc. This is not necessary with hyperbolic functions like sinh, cosh, and tanh.) I can't supply a specific answer for the second part, since your result will depend on the number of digits your calculator uses in its calculations.

5. $z' = 2b \sin(bx) \cos(bx)$

7. Suggestion—sketch a diagram consisting of a rectangle that is x cm wide by y cm high. A smaller rectangle lies inside it, one that is $x - 12$ cm by $y - 20$ cm. The area of the seed bed is then $A_{sb} = (x - a)(y - b)$ cm^2, with $a = 12$ and $b = 20$, and the total area is $A_t = xy$, with $A_t = 2000$. The idea in max-min problems is usually to reduce the problem to a single equation for the quantity to be optimized, as a function of a single variable. Here we can obtain A_{sb} in terms of x, namely

$$A_{sb} = (x - a)\left(\frac{A_t}{x} - b\right).$$

Differentiating yields

$$A'_{sb} = -b + \frac{A_t}{x} - \frac{(x - a)A_t}{x^2},$$

and setting that to zero gives $x = \pm\sqrt{abA_t}/b$, of which only the positive value is of interest. Thus $x = 34.641$ cm, from which $y = A_t/x$ or about 57.735 cm. The seed-bed area is then about 854.36 cm^2.

9. From the hint, $p = m - \delta(x - fn_t)$. Note as a check that at the start, when $x = fn_t$, $p = m$ as prescribed. Let A denote the total acorn production per ha per yr, so

$$A = xp = \left[\text{oaks} \cdot \left(\text{acorns oak}^{-1}\,\text{yr}^{-1}\right)\right] = x\,[m \div \delta(x - fn_t)].$$

The derivative dA/dx is zero when $x = (m + \delta fn_t)/2\delta$. That yields $x = 49$, only one more than the 48 that were there at the start. The acorn production density with 49 oaks would be 9604 kg ha^{-1} yr^{-1}. Acorn production density for the original 48 oaks would be 9600 kg ha^{-1}.

11.

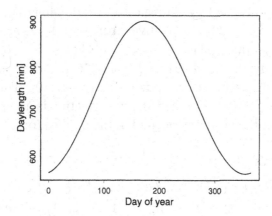

$$\frac{dL}{dt} = \frac{2b\pi}{c} \cos\left[\frac{2\pi(t+d)}{c}\right]$$

Daylength is increasing by about 0.5648497 min da^{-1} on Jan. 1, and decreasing by about 2.934913 min da^{-1} on Sept. 21.
$[L(0.5 + 0.005) - L(0.5 - 0.005)]/(2 * 0.005) = 0.5648497$, an excellent check.

13. Let L be the 3 mi along the road and x be the 1 mi distance from the road. Then the optimum target is a distance y mi south of the truck, with $y = L - x/\sqrt{3} \approx 2.4227$ mi. It would take you $t = (r/2) + (y/4)$ hours to reach the truck by that route, where $r^2 = x^2 + (L - y)^2$. Thus, $t \approx 1.18$ hr, just a little less than the 1.25 hr it would take by the right-angle route. Going straight for the truck would be slowest, at about 1.58 hr.

15. When the equation is simplified, the constant terms and those involving x^2 cancel out, and the original cubic results.

17. The series is $f(x) = (R + Sa + Ta^2 + (S + 2Ta)(x - a) + T(x - a)^2)$, which does simplify to the original quadratic.

19. The series is $f(x) \approx e + 2e(x-1) + (3/2)e(x-1)^2 + (2/3)e(x-1)^3$. Its fractional relative errors at the specified values of x are $-0.00001755649, -0.0002387804, -0.002819132, -0.02612107, -0.1636307, -0.5670925$, and -0.9438657, respectively.

21. $R(T) = \sigma\epsilon[(a^4 - T_s^4) + 4a^3(T - a) + 6a^2(T - a)^2 + 4a(T - a)^3 + (T - a)^4]$.

23. A. $p(z) \approx 0.3682701402 - 0.1473080561(z - 0.4) - 0.1546734589(z - 0.4)^2 + 0.06972581320(z - 0.4)^3$.

   B. $p(0) = 0.3989423$; $p(0.8) = 0.2896916$.

   C. $p(0) \approx 0.3979832$; Relative error: -0.002404107

   D. $p(0.8) \approx 0.2890616$; Relative error: -0.002174727

25. The series requested is $0\,(T - A)^0 + 4\,A^3\,(T - A)^1 + 6\,A^2\,(T - A)^2$. Numerically, the third term is 13,053,750, and the sum of the first two (just the second one, actually) is 513,447,500. The ratio of the two is about 0.025; whether the linearization is acceptable would depend on the accuracy needed in the application.

Chapter 3

1. A. $[(\log x)^2/2] + C$;   B. $[\log(x^2 + 1)/2] + C$;   C. 1;

   D. $[\exp(2t) - \exp(-2t) + 4t]/2$;   E. $e - 2$;   F. $1 - 3e^{-2}$

3. Simpson's rule yields $I \approx 2.906442$.

5. $36a + 144b + 648c$ m^3

7. For the linear equation, $A \approx -7.58190$ gb and $B \approx 5.387931$ gb yr^{-1}.

   Remember, keep lots of decimals in calculations, and round off only at the end. Roundoff error is an insidious problem, worse than most people seem to appreciate, and keeping a few extra digits can sometimes make a substantial difference in a final answer. Try doing the remaining calculations with the coefficients given here, and then again with $A \approx -7.6$ gb and $B \approx 5.4$ gb yr^{-1}. What differences do you see in the final answers? The roundoff effect is likely to be even greater with the exponential function in the next part.

   For the exponential equation, $a \approx 0.00152544$ gb and $b \approx 3.86155$ yr^{-1}.

   From these, we can calculate that the mean population over the time interval would have been 1.75 gb if the linear model were true, and 1.3953 gb if the exponential model were true. (*Now* we could reasonably round that latter answer to 1.4×10^9 people.) The rates of increase in the year 1732 would have been 5.3879 gb/yr and 4.7294 gb/yr under the two models.

9. This problem requires two separate applications of Simpson's 1/3 rule. As with many if not most realistic mass-balance problems, the units for various quantities are not consistent, and you have to settle on a consistent set. Here I'll use cubic meters for volume, and seconds for time. You may have chosen other units, but

if you dealt with them correctly, you should get the same overall results as are provided here.

The water volume added in the first day (two 12-hr periods) is (within the approximation resulting from integrating by Simpson's rule) 43200 m³, and the volume added in the second two 12-hour periods is 41184 m³. I'll leave you to add those to the original volume present, to find the amounts present at the end of each day.

With the units used here,

$$\text{hr} \cdot \frac{\text{s}}{\text{hr}} \cdot \frac{\text{m}^3}{\text{s}} = \text{m}^3.$$

11. 12.0047 kg. A unit check would account for $[\text{g da}^{-1}] \times [\text{da}] \times [\text{kg g}^{-1}]$, yielding kg as required.

13. Your exact result will depend on which three panels you apply the 3/8 rule to. Applying that rule to the 9th–11th panels and the 1/3 rule to the others yields a mean square root of 1.59 cm$^{1/2}$. Using a similar calculation yields an area of about 10.7 cm².

Chapter 4

1. The Maclaurin series is $e^t = 1 + t + t^2/2! + t^3/3! + t^4/4! + t^5/5! + \dots$. The derivatives, term by term, are 0, 1, $2t/2! = t$, $3t^2/3! = t^2/2!$, etc. Adding those derivatives gives you back the original series.

3. The equation is of the form $y' = a + by$, with $y = m$, $a = q_a c_A$, and $b = -[f + (q_a + q_b)/V]$. Thus, the general solution for that form applies. See Eqn 4.6, p. 93.

5. First find equations for the rise and fall of c_{in} with time. They would be $c_{in} = (C_{max}/7)t$ for $0 \le t \le 7$ days, and $c_{in} = C_{max}(2 - t/7)$ for $7 < t \le 14$ days. Define $(C_{max}/7) \stackrel{\text{def}}{=} \alpha$ for convenience. Then

$$\frac{dm}{dt} = qc_{in} - \frac{q}{V}m$$

is the rate equation for the insecticide mass in the tank for the first seven days. Rewrite that as

$$\frac{dm}{dt} = q\alpha t - \frac{q}{V}m.$$

For the second week, we have

$$\frac{dm}{dt} = qc_{in} - \frac{q}{V}m = qC_{max}\left(2 - \frac{t}{7}\right) - \frac{q}{V}m,$$

which, after a bit of algebra, can be rewritten as

$$\frac{dm}{dt} = 14q\alpha - q\alpha t - \frac{q}{V}m.$$

Now divide by both m' equations by V to obtain

$$\frac{dc}{dt} = \frac{q\alpha}{V}t - \frac{q}{V}c = \beta t + \gamma c, \text{ and}$$

$$\frac{dc}{dt} = \frac{14q\alpha}{V} - \frac{q\alpha}{V}t - \frac{q}{V}c = \delta - \beta t + \gamma c,$$

where $\beta = q\alpha/V$, $\gamma = -q/V$, and $\delta = 14q\alpha/V$. You can solve both of those using the second equation in the table on p. 107 of the notes. For the first week, you get

$$c(t) = -\frac{\beta\left(\gamma t + 1 - e^{\gamma t}\right)}{\gamma^2},$$

which, after replacing β and γ with their definitions becomes

$$c(t) = \frac{\alpha}{q}\left[qt - V + V\exp\left(-\frac{qt}{V}\right)\right].$$

The solution of the DE for the second week is

$$c(t) = \alpha(14 - t) + \frac{\alpha V}{q} + C_1 \exp\left(-\frac{qt}{V}\right),$$

where C_1 is the constant of integration. (You could obtain its value from the IC, if asked to do so.) The IC for that second equation is $c(7)$, as calculated from the first week's solution. If you plot the overall solution, you will see that the tank concentration never gets as high as C_{max}, and that it begins to drop at the point when c_{in} drops below c_{tank}. It does not begin to drop at the end of exactly 7 days.

7. $y_0 \exp(-3t)$; $(2/3)x^2 + y_0$; $y_0 \exp(-3.5t^2)$; and $-(1/3)\cos(3t) + (1/3) + y_0$.

9. See Eqn. 1.6, p. 9.

11. The DE is

$$\frac{dm}{dt} = rA - \frac{q}{V}m - fm.$$

This is another instance of our $y' = a + by$ form, with $y = m$, $a = rA$, and $b = -(f + q/V)$, so the solution again comes from Eqn 4.6, p. 93.

13. Because the derivative of $\log(cx)$ is $1/x$, then treating W/W_0 as cW, we see that the derivative of the U given is $0 + k/W$. That is clearly the same as the RHS of the DE. At $W = W(0)$, we have $U = U_0 + k\log(1) = U_0$, so both parts of the check are complete.

15. A↔iv; B↔iii; C↔i; D↔ii.

17.

$$\frac{dT}{dt} = \frac{dT}{dH}\frac{dH}{dt} = \frac{1}{C}\left[Qf - \frac{k}{S}(T - T_e)\right],$$

where $T_e = a + b\sin(rt)$. Each term has units of deg min^{-1}.

Chapter 5

1. The general analytic solution is that mass m reaches a given fraction m/m_0 of the initial mass at time

$$t = \frac{\log(m/m_0)}{\log(1/2)}H.$$

For ^{210}Pb, the useful range is 0.03176–219.25 yr. For ^{14}C, the range is 8.3285–57,502.6 yr. Thus, the ranges do overlap.

3. Ignoring new water coming in, the water *present* at some time $t = 0$ is lost according to $dw/dt = -(q/V)w$. Thus, the amount of that original water remaining at later times is $w = w_0 e^{-qt/V}$. Then $w/w_0 = 1/3$ when $t = -(V/q)\log(1/3)$. That turns out to be about 2.75 da.

5. $H = -(V/q)\log(1/2)$.

7. The integrating factor is $e^{0.2x}$, so $y = 7.5(e^{t/5} - 1) - 1.5x$.

9. $y = \sin(x) + (2/x)\cos(x) + (2/x^2)[\pi - \sin(x)]$.

Chapter 6

1. The volume equation is

$$\frac{dV}{dt} = q_{in}(t) - q_{out} - E(t); \quad V(0) = V_0,$$

and the one for PCB mass is

$$\frac{dB}{dt} = q_{in}c_{in} + k_1 D_a - \frac{q_{out}}{V}B - p_2 B; \quad B(0) = B_0,$$

where k_1 is the conversion factor from ng to pg, i.e., 10^3. A unit check here yields pg da^{-1} for each term. To convert the equation to one for concentration C when V changes with time is a messy process. Because $C = B/V$, to obtain dC/dt you must use the quotient rule. Another approach would be to solve this system for mass $B(t)$ and volume $V(t)$, and then to get the concentration at any time using $B(t)/V(t)$.

3.

$$\frac{d\rho_E}{dt} = r_E\rho_E\left(\frac{K_E - \rho_E}{K_E}\right) - \frac{k}{A_E}(\rho_E - \rho_P),$$

$$\frac{d\rho_P}{dt} = r_P\rho_P\left(\frac{K_P - \rho_P}{K_P}\right) + \frac{k}{A_P}(\rho_E - \rho_P).$$

k is the migration proportionality coefficient, with units of (rat da^{-1}) per rat m^{-2}), or m^2 da^{-1}.

5. $S' = -kSI$; $I' = kSI - fI$; and $R' = fI$. The ICs might be $S(0) = S_0$, $I(0) = I_0$, and $R(0) = R_0$, where at least the first two must be greater than zero if the epidemic is to occur. The units of k and f are [people^{-1} da^{-1}] and [da^{-1}] respectively.

7.

A. $$\frac{dm_B}{dt} = U - fm_B - k\left(\frac{m_B}{B} - p\frac{m_F}{F}\right);$$

$$\frac{dm_F}{dt} = k\left(\frac{m_B}{B} - p\frac{m_F}{F}\right);$$

B. $$\frac{dc_B}{dt} = \frac{U}{B} - fc_B - \frac{k}{B}(c_B - pc_F);$$

$$\frac{dc_F}{dt} = \frac{k}{F}(c_B - pc_F).$$

m_B, m_F, c_B, and c_F are the masses and concentrations of substance S in the blood and fat compartments, respectively, and k is the proportionality coefficient for the exchange between blood and fat, with units of kg da^{-1}.

Chapter 7

1. The analytic solution is

$$y = \left[0.7^{0.9} + 2.025(t^{1.2} - 1)\right]^{10/9},$$

so $y(1.4) \approx 1.841899$. Using Euler's method as specified yields 1.817347, which is about 1.3% too small.

3. What happens here is that the rate constants would be 365 times larger on a per-year basis than they are on a per-day basis, and this makes the slopes numerically that much larger. On the other hand, the values of h (and of $h/2$) are a factor of 365 times smaller, so each product of the form (time step) × (slope) remains constant. This exercise illustrates the general fact that you can change units in differential equations and still get the same results overall.

5. $H(0.1) \approx 5732.02$ kg and $C(0.1) \approx 4836.6$ kg. These large changes in just 1/10 year suggest that the calculations should probably be repeated with smaller time steps.

7. The error would be larger with (A) because with it, *both* the problems stated on p. 150 would occur, while with (B), the second problem (stated in the third bullet) does not apply.

Chapter 8

1. For Eqn 8.11, $C' = a\,(A \cosh ax + B \sinh ax)$ and $C'' = a^2\,(A \sinh ax + B \cosh ax)$, so $C'' = a^2 C$, as required. For Eqn 8.12, $C' = a\,(A\,e^{ax} - B\,e^{-ax})$ and $C'' = a^2\,(A\,e^{ax} + B\,e^{-ax})$, so again $C'' = a^2 C$, as required.

3. The equation is

$$\frac{d^2 T}{dz^2} = -\frac{M}{k} = -a,$$

where a is just a shorter name for M/k. You can solve this equation by separation, yielding $T(z) = A + Bz - a\,z^2/2$, where A and B are the constants of integration. Solving for those from the boundary conditions yields

$$T(z) = T_0 + \left(\frac{aL^2 - 2\,T_0 + 2\,T_L}{2L}\right) z - \left(\frac{a}{2}\right) z^2.$$

A quick check shows that this at least echoes the two boundary conditions.

5.

$$D\frac{d^2\rho}{dx^2} = -f\,\rho; \quad \rho(0) = \rho_A; \quad \rho(L) = \rho_B.$$

7. The equation is

$$\frac{d^2T}{dr^2} + \frac{2}{r}\frac{dT}{dr} = -\frac{q}{k}.$$

The solution is messy, but turns out to be

$$T(r) = \frac{-6\,kR_2T_2 - qR_2{}^3 + 6\,kR_1T_1 + qR_1{}^3}{6k\,(-R_2 + R_1)} -$$

$$\frac{R_1\,R_2\,\left(6\,k\,T_1 + qR_1{}^2 - 6\,k\,T_2 - qR_2{}^2\right)}{6rk\,(-R_2 + R_1)} - \frac{qr^2}{6k}$$

9. A. The annular ring in Fig.8.5, p. 179, from r to $r + \Delta r$ is a useful "control area" for deriving this equation. One obtains

$$\frac{d^2\rho}{dr^2} + \frac{1}{r}\frac{d\rho}{dr} = 0.$$

B. The solution is

$$\rho(r) = \frac{\rho_0\,\log\,(R_{outer}/r)}{\log\,(10\,R_{outer})}.$$

Note that it does give the correct values for ρ at $r = 0.1$ and at $r = R_{outer}$.

C. Thus, $\rho(5\ \text{km}) = 752.575$ lemmings m^{-2}.

Chapter 9

1. $x_2 \approx 0.00656122$ and $x_1 \approx 0.60946246$. You'll find that these check if you substitute them back into the original equations (as is always recommentd!). In the original matrix, the elements on the main diagonal are about double those on the "off" diagonal, so the system seems likely to be well determined. The determinant is about 576, which is more than the square of the largest element; this also suggests a well-determined system.

3. For $T_W = 100$ C, $T_2 \approx 73.45$ C, and for $T_W = 102.5$ C, $T_2 \approx 74.75$ C. When you solve a physical problem like this, it's always a good idea to think about whether your results seem reasonable. Here they seem to be, in the senses that the calculated temperatures start out near the water temperature and then drop in a reasonable progression toward the air temperature as we go from inside

to outside. Also, the inside wall is a little cooler than the water, and the outside wall is a little warmer than the air. For both water temperatures, we could apparently use the cheaper insulation, since $T_2 < 78$ C in both cases.

5. At equilibrium, this area would have about 24857 rabbits and 2729 foxes.

7. A.

$$A = \begin{pmatrix} \frac{q}{V_1} + f_1 & 0 & -\frac{q_f}{V_1} \\ \frac{q}{V_2} & -\left[\frac{q}{V_2} + f_2\right] & 0 \\ 0 & \frac{q}{V_3} & -\left[\frac{q}{V_3} + f_3\right] \end{pmatrix}; \quad B = \begin{pmatrix} \frac{q_i C_i}{V_1} \\ 0 \\ 0 \end{pmatrix}.$$

B.

$$A = \begin{pmatrix} 2.8 & 0 & -0.125 \\ 2 & -2.7 & 0 \\ 0 & 2 & -2.6 \end{pmatrix}; \quad B = \begin{pmatrix} 225 \\ 0 \\ 0 \end{pmatrix}.$$

C. The equilibrium ODW concentrations [g m^{-3}], are about 82.45, 61.08, and 46.98, respectively.

Chapter 10

General note: Don't forget that all the methods described in this book for finding roots are designed to find values of a variable that drive a function to **zero**.

1. To obtain Eqn 10.8, rearrange

$$x_{i+1} = x_i - \frac{x_i^2 - N}{2x_i}.$$

You should find $x \approx 4.123106$ when $N = 17$.

3. A. Because different combinations of parameters will arise in the future, it would be desirable to find an analytic solution. Unfortunately, such a solution is not available for this situation.

B. Here it turns out we *can* get an analytic solution. Set the two functions of time equal to one another, take the logs of both sides, and then solve for t. The result is that the two temperature differences would be equal when

$$t = \frac{\ln\left(\frac{\theta_{1,0}}{\theta_{2,0}}\right)}{b_1 - b_2}.$$

5. A.

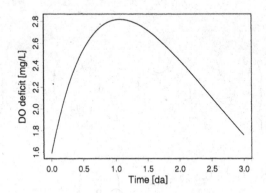

B. If you make the suggested plot, you'll see that deficit increases to more than 2 mg L^{-1} between about 0 and 0.3 da, and decreases back to 2 at between about 2.5-3 da. You could use any of the methods we've studied here to pin the times sought down to 0.16199656 da and 2.64859468 da. The problem statement calls for a range of *distances*, however, so we calculate how far water would move downstream in those times at a velocity of 1.5 ft s^{-1}; the results are 3.98 mi and 65.01 mi. This stream is in pretty bad shape!

7. In reservoir A, the concentration varies as $C_A(t) = C_{A0}e^{-t/\tau_A}$, with a similar equation for reservoir B. The numerical values are $C_{A0} = 80$, $\tau_A = 55000/2700$, $C_{B0} = 60$, and $\tau_B = 38000/700$. The units of the τ's are da^{-1}. When $C_B - C_A - 1 = 0$, $t \approx 10.0404$ da.

9. The concentration would first exceed 15 mg L^{-1} after about 1.6907 years. As you would see if you plotted the function, there is another root at about 5 years—that's when the concentration would decline below 15 mg L^{-1} again.

Chapter 11

1. We want the derivative of G with respect to N, treating P temporarily as a constant. It may help to rewrite the growth function as

$$G = \beta \left(\frac{K_n}{1 + ae^{-r_n N}} + b \right),$$

where β, temporarily treated as a constant, replaces the second major factor in the RHS of Eqn 11.1. This makes clear that we just need to differentiate the expression in the remaining parentheses with respect to N, then multiply the result by β. The constant b doesn't contribute to the derivative either, and we can factor out K_N. Thus, we need to differentiate $(1 + ae^{-r_n N})^{-1}$. We obtain $-(1 + ae^{-r_n N})^{-2}$, times the derivative of the $ae^{-r_n N}$ term, which is $-ar_n e^{-r_n N}$. Cancelling the two minus signs and replacing the K_N and β then yields the derivative in Eqn 11.2.

3. I would work with a control volume (CV) that is H mm thick, by Δx cm in the x direction, by Δy cm in the y direction. Because other length dimensions are all expressed in cm, convert H to cm too. The equation sought is

$$\frac{\partial T}{\partial t} = \frac{k}{c_p \rho} \left[\frac{\partial^2 T}{\partial x^2} + \frac{\partial^2 T}{\partial y^2} \right] + \frac{S}{c_p \rho H} - \frac{b(T - \theta)}{c_p \rho H}.$$

5.

$$\frac{\partial N}{\partial t} = m \frac{\partial^2 N}{\partial x^2} + rN \frac{K - N}{K}.$$

7.

$$\frac{\partial c}{\partial t} = -\frac{q}{WH} \frac{\partial c}{\partial x} - pc.$$

The initial condition is a specification of c at time zero, for all $0 \le x \le L$.

9. $\Delta t = 0.03125$ hr, or a bit under 2 min. In the table below, the columns for $z = 0$ and $z = 12$ cm (the given boundary conditions) have been removed so the rest will fit the page width.

	Depth z [cm]						
	1.5	3.0	4.5	6.0	7.5	9.0	10.5
t	Temperature [deg C]						
0	9.8	11.3	11.9	11.1	8.9	6.0	3.2
Δt	9.5	10.85	11.2	10.4	8.55	6.05	3.9
$2\Delta t$	9.025	10.35	10.625	9.875	8.225	6.225	3.925
$\cdots$			$\cdots$				

11. Each term in this result has units of kg biomass ha^{-1} da^{-1}:

$$\frac{\partial B}{\partial t} = D\left(\frac{\partial^2 B}{\partial r^2} + \frac{1}{r}\frac{\partial B}{\partial r}\right) + G.$$

Appendix C

List of Applications

Bibliography

Abramowitz, M. and I.A. Stegun, (eds.) 1972. *Handbook of Mathematical Functions with Formulas, Graphs, and Mathematical Tables.* Dover Publications, New York.

Arfken, G. 1970. *Mathematical Methods for Physicists.* Academic Press, New York.

Beissinger, S.R. and M.I. Westphal. 1998. On the use of demographic models of population viability in endangered species management. *Journal of Wildlife Management* 62: 821–841.

Carslaw, M.S. and J.C. Jaeger. 1947. *Conduction of Heat in Solids.* Oxford University Press, London.

Caswell, Hal. 2001. *Matrix Population Models. Construction, Analysis, and Interpretation,* 2nd ed. Sinauer Associates, Inc., Sunderland, MA.

Chaston, I. 1971. *Mathematics for Ecologists.* Butterworths, London.

Chen, L. 1995. Testing the mean of skewed distributions. *Journal of the American Statistical Association* 90: 767-772.

Derrick, W.R. and S.I. Grossman. 1981. *Elementary Differential Equations with Applications.* Addison-Wesley Publishing Co., Reading, MA.

Edwards, C.H. and D.E. Penney. 1982. *Calculus and Analytic Geometry.* Prentice-Hall, Inc, Englewood Cliffs, NJ.

Geiger, R. 1966. *The Climate Near the Ground.* Cambridge, MA, Harvard University Press.

Greene, W.H. 1993. *Econometric Analysis.* Macmillan Publishing Company, New York.

Grossman, S.I. and J.E. Turner. 1974. *Mathematics for the Biological Sciences.* Macmillian Publishing Company, New York.

Haefner, J.W. 2005. *Modeling Biological Systems.* 2nd Ed. Springer Science+Business Media, New York.

Hanselman, D. and B. Littlefield. 2001. *Mastering MATLAB 6.* Prentice Hall, Upper Saddle River, NJ.

Higham, D.J. and N.J. Higham. 2000. *MATLAB Guide.* Society of Industrial and Applied Mathematics (SIAM), Philadelphia.

Holmes, E.E., M.A. Lewis, J.E. Banks, and R.R. Veit. 1994. Partial differential equations in ecology: spatial interactions and population dynamics. *Ecology* 75: 17–29.

Horn, H.S. 1976. Succession. In R.M. May, ed. *Theoretical Ecology.* W.B. Saunders Co., Philadelphia. pp. 187–204.

Kermack, W.O. and A.G. McKendrick. 1927. Contributions to the mathematical theory of epidemics. *Proceedings of the Royal Society A* 115:700-721.

Kofler, M. 1997. *Maple. An Introduction and Reference.* Addison-Wesley, Reading, MA.

Kreith, F. 1973. *Principles of Heat Transfer.* Intext Educational Publishers, New York.

Lide, D.R. 2005. *CRC Handbook of Chemistry and Physics,* 86th Ed. CRC Press, Boca Raton, FL.

List, R.J. 1949. *Smithsonian Meteorological Tables.* Smithsonian Institution Press, Washington, DC.

Mesterton-Gibbons, M. 1995. *A Concrete Approach to Mathematical Modelling.* John Wiley & Sons, New York.

Maki, D.P. and M. Thompson. *Mathematical Models and Applications.* Prentice-Hall, Inc., Englewood Cliffs, NJ.

Murphy, G.M. 1960. *Ordinary Differential Equations and Their Solutions.* D. Van Nostrand Co., Princeton, NJ.

Nobel, P.S. 1991. *Physicochemical and Environmental Plant Physiology.* Academic Press, San Diego.

Parkhurst, D.F. 1994. Diffusion of CO_2 and other gases inside leaves. *The New Phytologist* 126: 449-479.

Porter, W.P., D.F. Parkhurst, and P.A. McClure. 1986. Critical radius of endotherms. *American Journal of Physiology 250 (Regulatory, Integrative, and Comparative Physiology)* 19: R699-R707.

Press, W.H., S.A. Teukolsky, W.T. Vetterling, and B.P. Flannery. 1992. *Numerical Recipes in FORTRAN.* Cambridge University Press, New York.

Schey, H.M. 1973. *Div, Grad, Curl, and All That.* W.W. Norton & Co., New York.

US EPA. 1996. *Soil Screening Guidance: Technical Background Document.* Washington, DC 20460, US Environmental Protection Agency, Office of Research and Development. 9355.4-17A; EPA/540/R-95/128; PB96-963502.

van Wijk, W.R. (ed.). 1963. *Physics of Plant Environment*, North-Holland Publishing Co., Amsterdam.

Index